저탄소 플랜트 개념설정과 경제성 및 사업타당성 분석

고동현 ◆ 著

에듀컨텐츠·휴피아

목 차

제1장. 개 요 ·· 3

제2장. 탄소저감 & 탄소중립 ·· 11

제3장. 탄소중립 에너지 기술 1 ·· 53

제4장. 탄소중립 에너지 기술 2 ·· 73

제5장. 화학공장 Process 설계 1 ······································ 103

제6장. 화학공장 Process 설계 2 ······································ 121

제7장. Flow Diagram ·· 145

제8장. 화학 공정 반응기 ·· 175

제9장. 반응기 설계 1 ·· 209

제10장. 반응기 설계 2 ·· 243

제11장. 열교환기 ··· 253

제12장. 분리장치 ··· 271

제13장. 투자비 및 생산비용/매출 ······································ 291

제14장. 생산비용 및 매출 추정 ··· 319

제15장. Chemical Eng. Design Project - A case study approach ··· 357

에듀컨텐츠·휴피아
CH Educontents·Huepia

저탄소 플랜트 개념설정과 경제성 및 사업타당성 분석

에듀컨텐츠 휴피아

제1장
개 요

저탄소를 향해 가야 하는 이유

This month is the planet's hottest on record by far – and hottest in around 120,000 years, scientists say

By Laura Paddison, CNN
Updated 4:57 PM EDT, Thu July 27, 2023

The heat in July has already been so extreme that it is "virtually certain" this month will break records "by a significant margin," the European Union's Copernicus Climate Change Service and the World Meteorological Organization said in a report published Thursday.

We have just lived through the hottest three-week-period on record – and almost certainly in more than a hundred thousand years.

지난 133년(1880~2012년)간 지구의 평균기온은 0.85℃ 상승
우리나라는 전세계 평균대비 2배 이상 빠르게 온도 상승

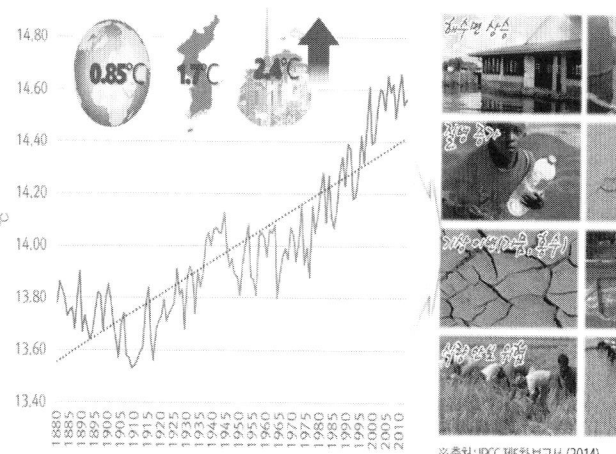

※출처: IPCC 제5차보고서 (2014)
- IPCC : Intergovernmental Panel on Climate Change
  (기후 변화에 관한 정부간 협의체, 1988년 설립)

"2025~2095년 사이 대서양 해류 멈춘다"
코펜하겐대 연구진 연구결과 발표

지구온난화로 빙하 녹아 염도차 줄어
적도~북미~북극 순환 해류 약해져
북미엔 극한추위, 적도 인근엔 폭염

올 여름 폭염, '탄소배출 결과' 연구도
"지구온난화 없었으면 안 일어났을 것"

CNN Report Link

지구 온도 조절 '컨베이어벨트' 멈춘다

지구 열 순환을 돕는 대서양 해류가 이르면 오는 2025년 멈춰설 수 있다는 연구 결과가 나왔다. 금세기 안에 이러한 현상이 일어나지 않을 것이라는 기존 과학계의 관측보다 시기가 훨씬 앞당겨진 것이다. 북미·유럽에는 혹독한 추위가 찾아오고 열대 지방은 더 더워지는 기후 재앙이 발생할 수 있다는 우려도 나온다.

"지구온난화 없었으면 폭염도 없었을 것"

이날 세계기상기여도그룹은 이달 들어 북반구에서 발생한 폭염이 19세기 후반부터 시작된 화석연료 배출로 인해 지구의 평균 기온이 화씨 2도 올라간 결과라는 연구 보고서를 발표했다. 이들은 기후온난화가 발생하지 않은 12가지 기후 모델을 가상 실험한 결과 이번 달 미국·멕시코·남부 유럽에 발생한 폭염은 지구 온난화가 없었다면 사실상 일어나지 않았을 것이라고 결론내렸다.

온실 가스 및 지구 온난화

CO_2	CH_4	N_2O	HFCs	PFCs	SF_6
• 화석연료 연소 • 보일러/차량 등	• 화석연료 연소 • 유기물 분해 • 반추동물 트림 등	• 화석연료 연소 • 질소질 비료 사용 • 석탄 채광 등	• 냉장고, 에어컨 냉매 등	• 전자제품, 도금산업, 반도체 제조시 세척 등	• 전기제품, 변압기 등의 절연가스 등

온실가스 (greenhouse gases, GHGs) 는 지구의 지표면에서 우주로 발산하는 적외선 복사열을 흡수 또는 반사하여 지구 표면의 온도를 상승시키는 역할을 하는 특정 기체를 의미

"온실가스"란 적외선 복사열을 흡수하거나 재방출하여 온실효과를 유발하는 가스 상태의 물질로서 법 제2조제9호에서 정하고 있는 이산화탄소(CO_2), 메탄(CH_4), 아산화질소(N_2O), 수소불화탄소(HFCs), 과불화탄소(PFCs) 또는 육불화황(SF_6) 등을 말하며 수소불화탄소(HFCs)와 과불화탄소(PFCs)에 대한 세부사항은 별표 3과 같다.

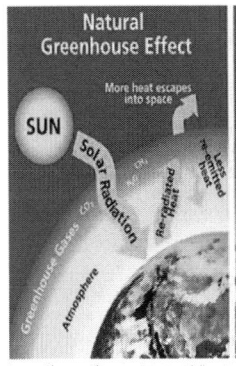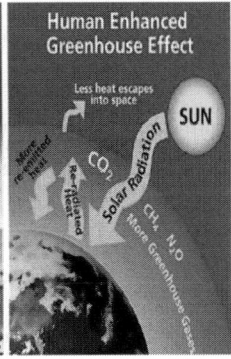

https://www.c2es.org/content/climate-basics-for-kids/

전 지구적 온실 가스 배출량

1750~2011년 기간 약 2,040억톤의 이산화탄소가 인위적으로 배출
1970년 이래로 화석연료 연소, 시멘트 생산 등으로 인한 누적 배출량은 3배로 증가

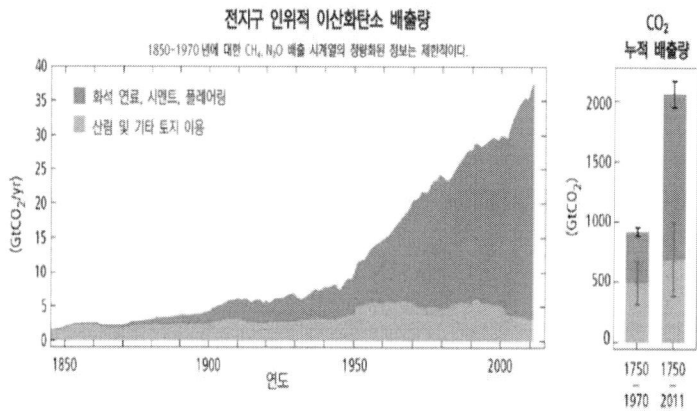

※ 출처. 제2차 기후변화 기본계획(2019, 관계부처 합동)

화학 반응

1) A + B → C
2) A + B → C + D   } Ideal Case

In Reality

➢ A + B → $_A$ + $_B$ + C + D + $_E$ + $_F$ + · · ·

Considerations : G-L-S, T, P,
　　　　　　　　Physical properties etc.

예 : Propylene to Acrylic Acid

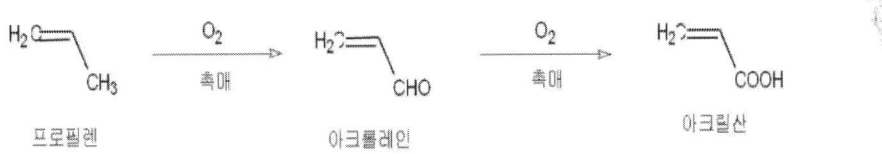

- 이론적으로는 H_2O만 생성

However,

→ CO, CO_2, Acetic acid, Propionic acid, · · · · · ·

Process Flow Diagram (예시)

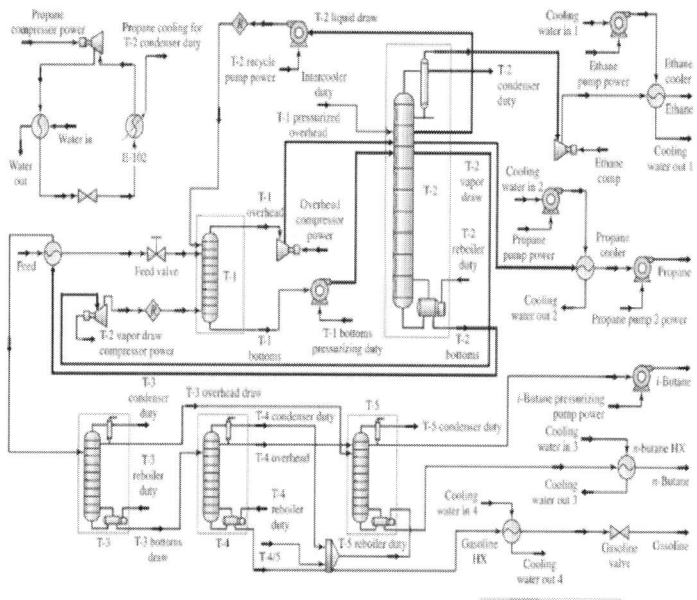

Plant Construction

Table 1-1 Typical design steps for chemical and biochemical processes

1. Recognize a societal or engineering need.
   a. Make a market analysis if a new product will result.
2. Create one or more potential solutions to meet this need.
   a. Make a literature survey and patent search.
   b. Identify the preliminary data required.
3. Undertake preliminary process synthesis of these solutions.
   a. Determine reactions, separations, and possible operating conditions.
   b. Recognize environmental, safety, and health concerns.
4. Assess profitability of preliminary process or processes (if negative, reject process and create new alternatives).
5. Refine required design data.
   a. Establish property data with appropriate software.
   b. Verify experimentally, if necessary, key unknowns in the process.
6. Prepare detailed engineering design.
   a. Develop base case (if economic comparison is required).
   b. Prepare process flwsheet.
   c. Integrate and optimize process.
   d. Check process controllability.
   e. Size equipment.
   f. Estimate capital cost.
7. Reassess the economic viability of process (if negative, either modify process or investigate other process alternatives).
8. Review the process again for environmental, safety, and health effects.
9. Provide a written process design report.
10. Complete the final engineering design.
    a. Determine equipment layout and specifications.
    b. Develop piping and instrumentation diagrams.
    c. Prepare bids for the equipment or the process plant.
11. Procure equipment (if work is done in-house).
12. Provide assistance (if requested) in the construction phase.
13. Assist with start-up and shakedown runs.
14. Initiate production.

- 정유 공장 건설 사례 동영상

경제성, 사업 타당성이란?

- 회사의 주요 기능 : 소유주와 주식 보유자에게 장기간의 이익을 최대로 제공하는 것

❖ 이익을 극대화하고 장기화 하기 위해서

- 돈과 자본 투하의 적절성을 평가
- 자본에 대한 시간적 가치를 평가
- 설비 등 물리적 자산에 대한 가치 및 시간흐름에 따른 평가
- 설계의 최적화
- 조업의 최적화
- ...

강의 계획

교재 : Plant Design and Economics for Chemical Engineers
   Max S. Peters 외 2인, McGraw-Hill, 2003

부교재 1 : Unit Operations of Chemical Engineering,
   Warren L. MaCabe 외 2인,
   McGraw-Hill, 2008(7th)

부교재 2 : 탄소중립과 에너지 기술, 윤양일,
   전남대학교 출판문화원, 2022

강의 계획

- 2~3주 : 탄소 저감/중립 및 저탄소 기술
- 4주 : 공장 건설을 위한 진행 단계에 대한 이해
- 5주 : 공정 및 장치 설계를 위한 과정과 주요 사항에 대한 이해
- 6~7주 : 반응기 및 이송 장치에 대한 이해, 설계와 비용
- 8주 : 열전달 장치에 대한 이해, 설계와 비용
- 9주 : 분리 및 정제 장치에 대한 이해, 설계와 비용
- 10~11주 : 경제성 및 사업타당성 평가 – 투자비 산출
- 12~13주 : 경제성 및 사업타당성 평가 – 돈의 시간 가치
- 14주 : 경제성 및 사업타당성 평가 – 수익성/대안 비교
- 15주 : 경제성 및 사업타당성 평가 – 실제 사례

강 의 노 트

1.

2.

3.

제2장

탄소저감 & 탄소중립

화학반응의 종류

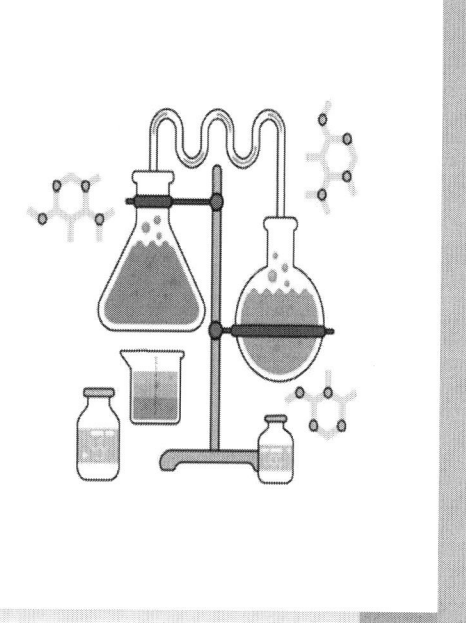

기본 화학 반응

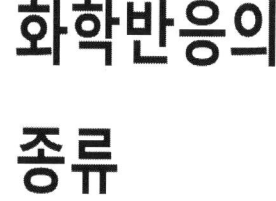

- 합성 : Na + Cl ⟶ NaCl
- 분해 : H_2O ⟶ H_2 + $1/2O_2$
- 치환 : $2Ag(NO_3)$ + Cu ⟶ $Cu(NO_3)$ + 2Ag
- 이중 치환 : HCl + NaOH ⟶ NaCl + H_2O

현재의 석유화학공장 운전을 위해서...

- 다양한 원료 (주로 Crude Oil에서)
- 열원은 주로 화석연료를 이용해서
- 고압은 전기를 이용해서 (전기는 주로 화석연료에서)
- 부산물 또는 폐기물의 안전한 처리

기후 변화의 원인

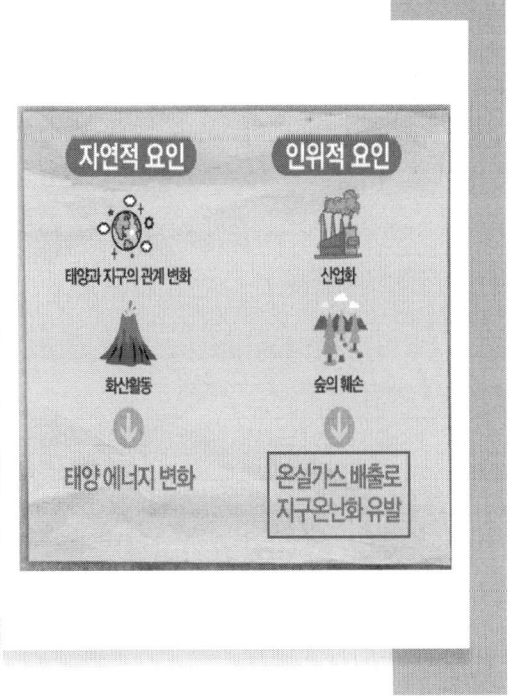

Global Warming Is Not A New Theory
It's the confirmation of a prediction

- 1890s – Nobel Prize winner Svante Arrhenius theorized about a warming climate due to the burning of coal.

- 1938 – Guy Stewart Callendar asserted that warming of the 19th century forward was due to a rise in CO_2.

- 1965 – Roger Revelle: "By the year 2000, the increase in atmospheric CO_2 ...may be sufficient to produce measurable and perhaps marked change in climate"

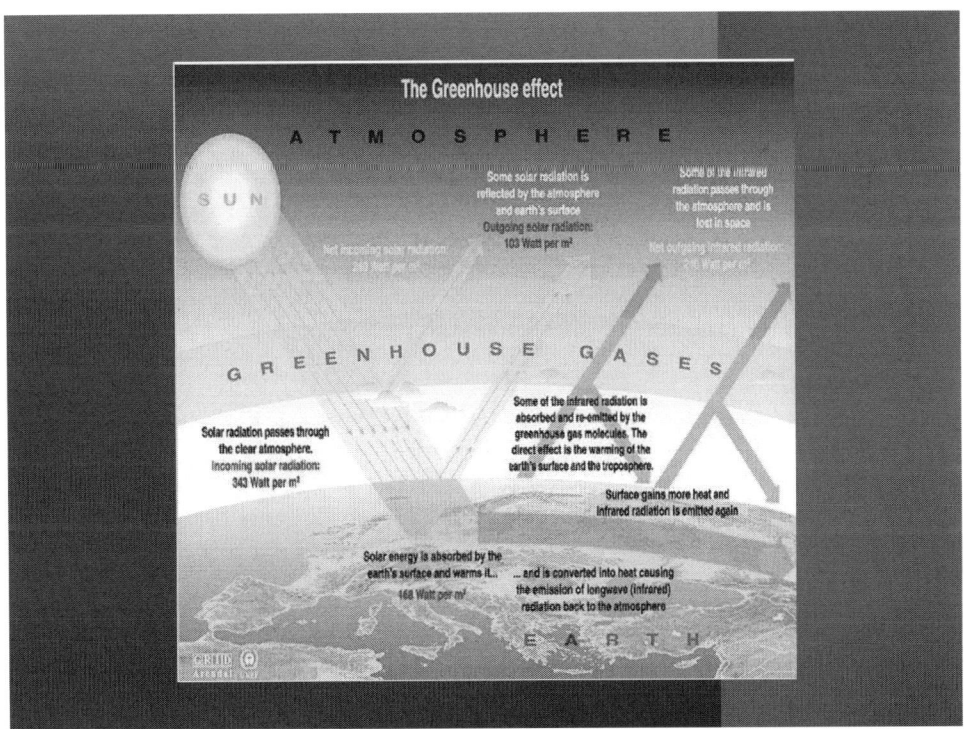

온실 가스 및 지구 온난화

CO₂	CH₄	N₂O	HFCs	PFCs	SF₆
• 화석연료 연소 • 보일러/차량 등	• 화석연료 연소 • 유기물 분해 • 반추동물 트림 등	• 화석연료 연소 • 질소질 비료 사용 • 석탄 채광 등	• 냉장고, 에어컨 냉매 등	• 전자제품, 도금산업, 반도체 제조시 세척 등	• 전기제품, 변압기 등의 절연가스 등

온실가스(greenhouse gases, GHGs)는 지구의 지표면에서 우주로 발산하는 적외선 복사열을 흡수 또는 반사하여 지구 표면의 온도를 상승시키는 역할을 하는 특정 기체를 의미

"온실가스"란 적외선 복사열을 흡수하거나 재방출하여 온실효과를 유발하는 가스 상태의 물질로서 법 제2조제9호에서 정하고 있는 이산화탄소(CO2), 메탄(CH4), 아산화질소(N2O), 수소불화탄소(HFCs), 과불화탄소(PFCs) 또는 육불화황(SF6) 등을 말하며 수소불화탄소(HFCs)와 과불화탄소(PFCs)에 대한 세부사항은 별표 3과 같다.

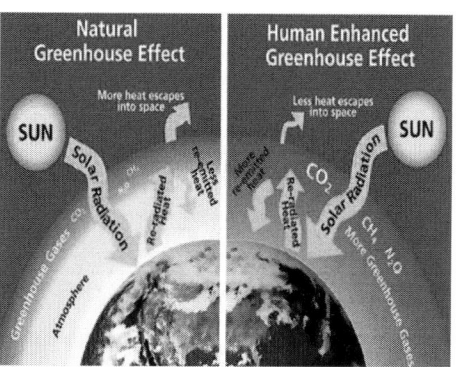

https://www.c2es.org/content/climate-basics-for-kids/

온실 가스의 주요 특징

온실가스 종류	지구 온난화 지수 (GWP*)	대기 체류시간 (년)	전 세계 총 배출량 (%)	온난화 총기여도(%)	주요 배출원
CO₂ (이산화탄소)	1	100~200	≒ 73	55	화석연료 연소, 차량/보일러
CH₄ (메탄)	21	10	≒ 19	15	화석연료 연소, 유기물 & 농/축산물 분해
N₂O (아산화질소)	310	100~150	≒ 5	6	화석연료 연소, 질소비료
HFCs (수소불화탄소)	140~11,700	-			냉매, 발포제, 반도체공정
PFCs (과불화탄소)	6,500~9,500	2,000~50,000	≒ 3	24	냉동기, 반도체 세정제
SF₆ (육불화황)	23,900	3,200			변압기, 절연체

*GWP : Global Warming Potential, 지구 온난화에 기여하는 정도
- IPCC (Intergovernmental Panel on Climate Change) 2차 평가보고서, 1995

전 세계 온실 가스 배출량 (1990~2019)

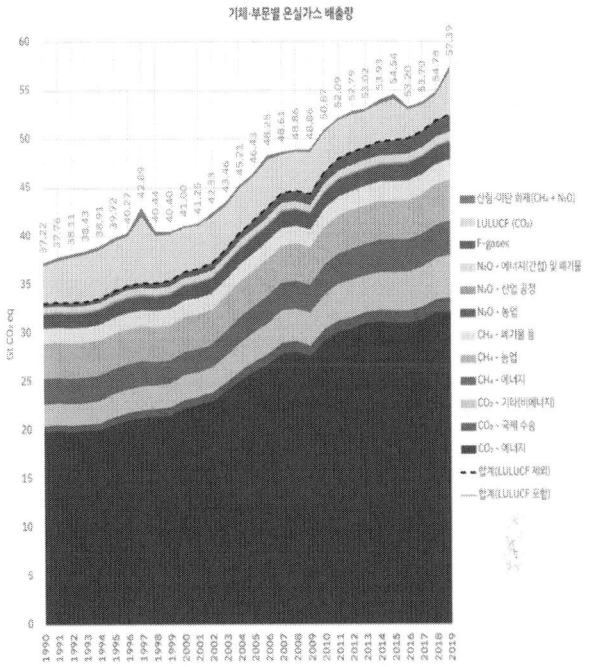

Source : Netherlands Environmental Agency

전 세계 국가별 온실 가스 배출량 (1990~2019)

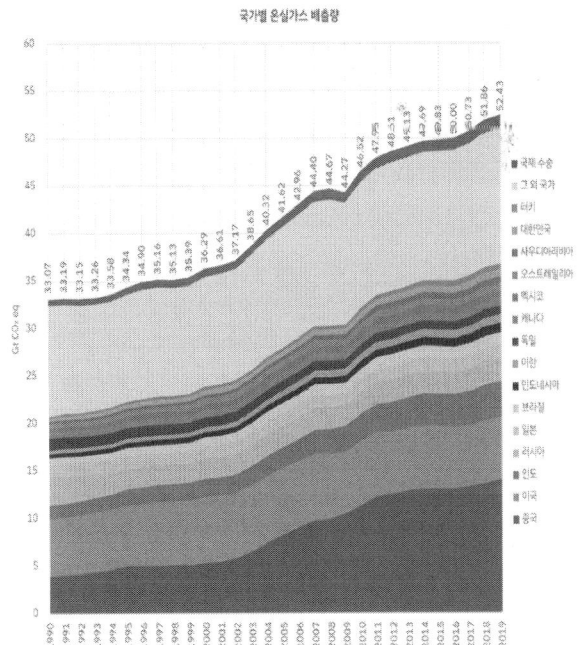

Source : Netherlands Environmental Agency

전 세계 부문별
온실 가스 배출량
비중(2016)

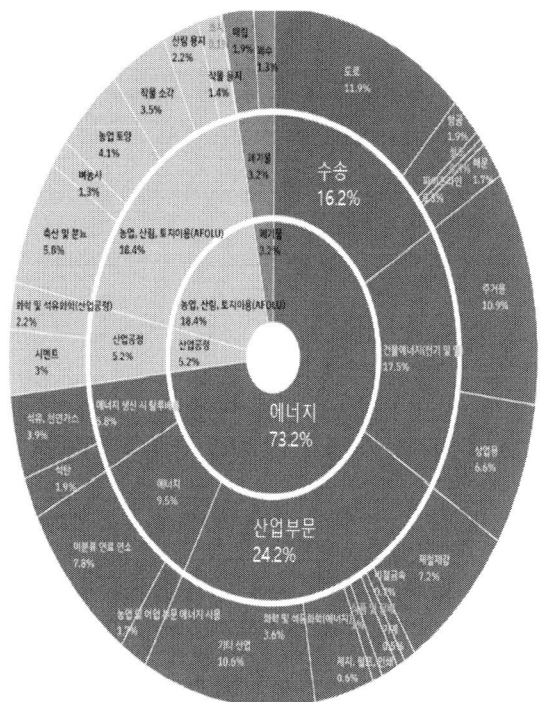

Source : http://ourworlddata.org/emissions-by-sector

전 세계

- 육지-해양 표면 온도 편차
- 평균 해수면 변화
- 평균 온실가스 농도 변화

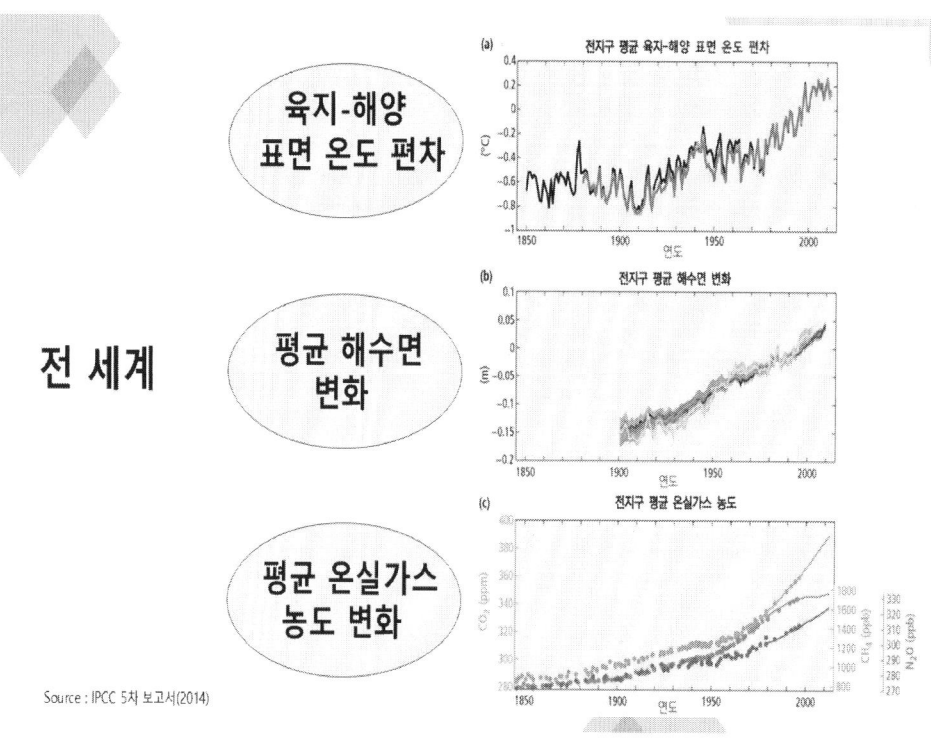

Source : IPCC 5차 보고서(2014)

전 세계 인위적 GHG 배출량

1970-2010년 연간 총 인위적 온실가스(GHG) 배출량

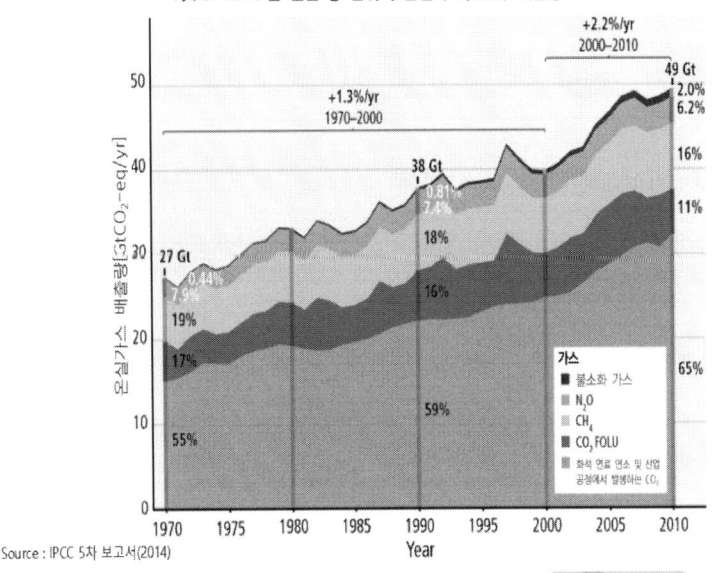

Source : IPCC 5차 보고서(2014)

전 지구적 CO_2 배출량

1750~2011년 기간 약 2,040억톤의 이산화탄소가 인위적으로 배출
1970년 이래로 화석연료 연소, 시멘트 생산 등으로 인한 누적 배출량은 3배로 증가

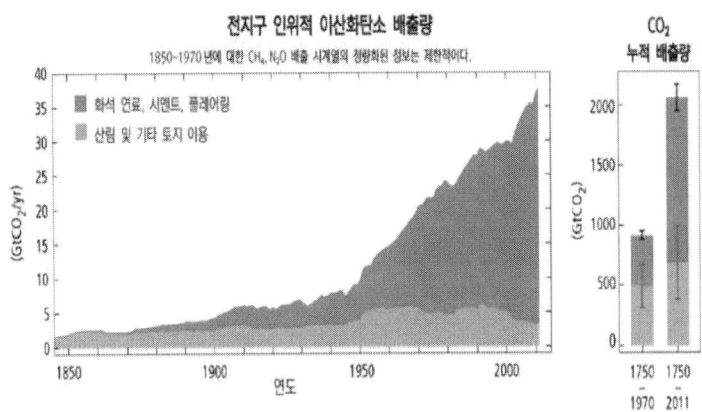

※ 출처. 제2차 기후변화 기본계획(2019, 관계부처 합동)

전 지구 기온 이상 변화 (1880년 기상관측 이후, 1940년 기준=13.6℃)

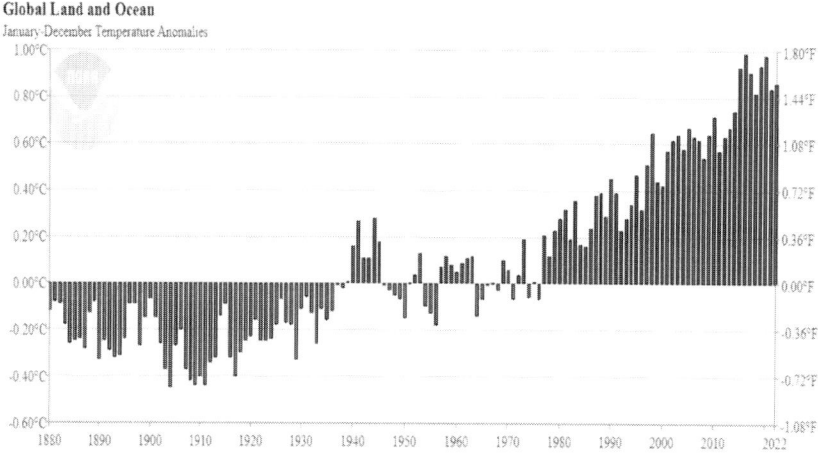

Source : National Centers for Environmental Information,
Annual 2022 Global Climate Report

전 지구 온도 변화 추이

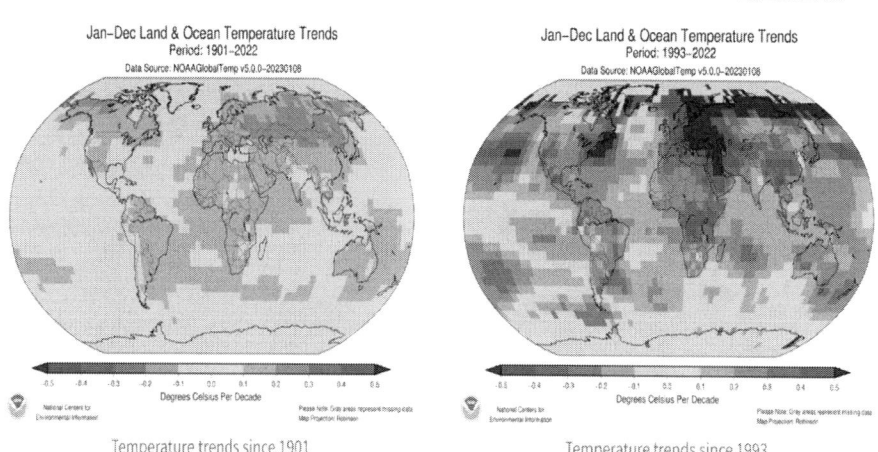

Source : National Centers for Environmental Information,
Annual 2022 Global Climate Report

"2025~2095년 사이 대서양 해류 멈춘다"
코펜하겐대 연구진 연구결과 발표

지구온난화로 빙하 녹아 염도차 줄어
적도~북미~북극 순환 해류 약해져
북미엔 극한추위, 적도 인근엔 폭염

올 여름 폭염, '탄소배출 결과' 연구도
"지구온난화 없었으면 안 일어났을 것"

지구 온도 조절 '컨베이어벨트' 멈춘다

지구 열 순환을 돕는 대서양 해류가 이르면 오는 2025년 멈춰설 수 있다는 연구 결과가 나왔다. 금세기 안에 이러한 현상이 일어나지 않을 것이라는 기존 과학계의 관측보다 시기가 훨씬 앞당겨진 것이다. 북미·유럽에는 혹독한 추위가 찾아오고 열대 지방은 더 더워지는 기후 재앙이 발생할 수 있다는 우려도 나온다.

"지구온난화 없었으면 폭염도 없었을 것"

이날 세계기상기여도그룹은 이달 들어 북반구에서 발생한 폭염이 19세기 후반부터 시작된 화석연료 배출로 인해 지구의 평균 기온이 화씨 2도 올라간 결과라는 연구 보고서를 발표했다. 이들은 기후온난화가 발생하지 않은 12가지 기후 모델을 가상 실험한 결과 이번 달 미국·멕시코·남부 유럽에 발생한 폭염은 지구 온난화가 없었다면 사실상 일어나지 않았을 것이라고 결론내렸다.

지난 133년(1880~2012년)간 지구의 평균기온은 0.85°C 상승
우리나라는 전세계 평균대비 2배 이상 빠르게 온도 상승

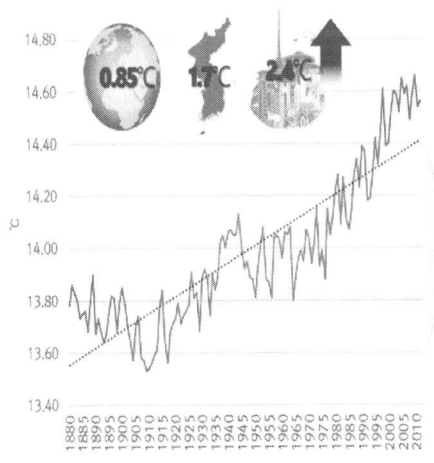

※ 출처: IPCC 제5차보고서 (2014)
- IPCC ; Intergovernmental Panel on Climate Change
  (기후 변화에 관한 정부간 협의체, 1988년 설립)

우리나라의 온실가스 배출량

[2018년 총 배출량은 7.2억톤, 이산화탄소배출량 세계 8위
 - 세계 1위는 중국, 2위는 미국, 3위 인도의 순]

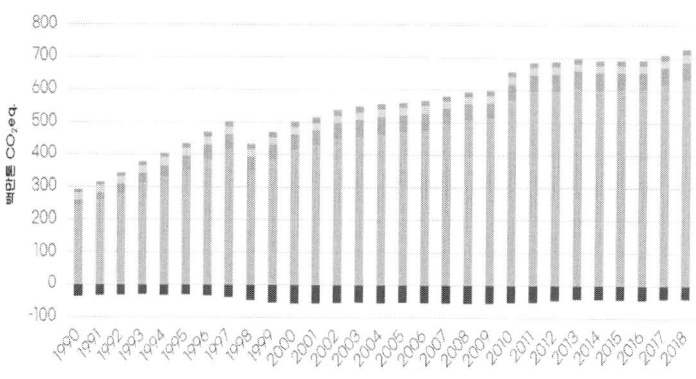

분야별 온실가스 배출량 및 흡수량(1990~2018)

■에너지 ■산업공정 ■농업 ■LULUCF ■폐기물

[출처] 2019 국가온실가스 인벤토리보고서

UN기후변화협약의 태동

- 1968년 로마클럽 아우렐리오 페체이
- 1972년 로마클럽 "성장의한계"
- 1972년 UN인간환경회의 "지속가능성"
- 1987년 세계기상회의 "기상이변"
- 1988년 정부간기후변화패널 "IPCC"
- 2015년 파리협정 "채택"
  * 제1차 평가보고서('90) → 유엔기후변화협약(UNFCCC) 채택('92)
  * 제2차 평가보고서('95) → 교토의정서 채택('97)
  * 제4차 평가보고서('07) → 노벨평화상 수상(엘 고어 공동 수상)
  * 제5차 평가보고서('14) → 파리협정 채택 '15)
- UN환경개발회의 "IPCC 1차보고서"
- 2005년 교토의정서 "발효"
- 1997년 교토의정서 "채택"
- 1994년 기후변화협약 "발효" 154개국
- 1992년 기후변화협약 "채택"

 * '01 미국 탈퇴

- IPCC : Intergovernmental Panel on Climate Change (기후 변화에 관한 정부간 협의체)
- UNFCCC : United Nations Framework Convention on Climate Change (유엔기후변화협약)

UN기후변화협약

■ 기후변화협약은 지구온난화 및 기후변화에 대응하기 위한 기본틀(Frame) 구축
■ 기후시스템과의 위험한 인위적 간섭 방지를 위한 온실가스 농도 안정화

기후변화협약의 의무 조항(4조) **"공동의 차별화된 책임의 원칙"**

구분	개도국 의무(4.1조)	선진국 의무(4.2조)
목적	온실가스 농도의 안정화 수준은 생태계가 적응하고, 농업생산이 위협받지 않으며, 경제가 지속가능한 방식으로 추구하기에 충분한 시계(time-frame)내에 달성	
대상국가	모든 가입국 (현 196개국)	선진국 (당시 24개 OECD국가, 11개 동구 유럽국가)
온실가스 통계 작성&제출	이산화탄소 등 온실가스 배출량 및 흡수량	좌동
온실가스 감축목표	없음	1990년 수준으로 2000년까지 온실가스 배출량 동결
국가전략 추진	기후변화 대응, 적응 위한 국가프로그램 수립, 시행, 공표	동결목표 달성을 위한 국가정책 채택
공동협력 사항	- 온실가스 감축기술 개발보급확산 - 흡수원 보호 및 증진 - 국가정책에 기후변화 반영	- 개도국에 대한 기술이전(4.5조) - 개도국에 대한 재정지원(4.3조) - 국가간 경제/행정수단의 통합적 추진 가능

교토의정서

■ 교토의정서 1차 공약기간은 2008년부터 2012년까지로 (이후 2020년까지 연장)
■ 38개 선진국이 1990년을 기준으로 온실가스 감축 의무

교토의정서의 의무 조항

구분	개도국 의무(10조)	선진국 의무(2,3조)
공약기간	1차, 2008년 ~ 2012년	
대상국가	38개 선진국 (OECD국가, 동구 유럽국가)	
온실가스 감축목표	없음	- 2008~2012년에 6대 온실가스를 1990년 수준 대비 6~8% 감축 - 의무 감축 및 법적 강제력
국가전략 추진	기후변화 대응, 적응 위한 국가 프로그램 수립, 시행	목표 달성을 위한 국가 정책 및 조치 채택
신규 제도 도입	- 배출권거래제(17조) - 공동이행제(6조) - 청정개발체제(CDM)(12조) - 공동목표설정(4조)	
기타	2012년 이후 의무에 대한 협상 개시 조항(3조 9항)	

교토의정서_CDM
Clean Development Mechanism

청정개발체제	목적	Non-Annex I 국가의 지속가능한 개발에 기여함과 동시에 Annex I 국가(선진국)가 온실가스 감축의무를 비용 효과적으로 달성하도록 도움
	대상	Annex I 국가가 Non-Annex I 국가에 기술 및 자본을 투자하여 온실가스 감축 실적을 인정받음
	주관기관	UNFCCC CDM 집행위원회 (Executive Board, EB)
	크레디트	CERs (Certified Emission Reductions)
	진행절차	사업계획 → 타당성평가 → 승인 및 등록 → 모니터링 → 검증 및 인증 → CERs 발행
	사업 인정 기간	Option 1 : 10년 (갱신 불가능) Option 2 : 7년 (갱신 가능)

교토의정서 제12조에 정의되어 있는 청정개발체제는 부속서 I국가(선진국)가 비부속서 I국가(개발도상국)에 온실가스 감축사업 실행을 위한 기술 및 자금을 지원하여 달성한 실적을 부속서 I국가(선진국)에 할당된 감축목표 달성에 활용할 수 있도록 하는 제도입니다. CDM 사업을 통하여 선진국은 감축목표 달성에 사용할 수 있는 온실가스 감축량을 얻고, 개발도상국은 선진국으로부터 기술과 재정지원을 받음으로써 자국의 지속가능한 개발에 기여할 수 있습니다.

교토의정서 이후

2002 미국 불참 선언
미국 정부 교토의정서 1차의무 불참 선언
- 미국 : 주요개도국 불참과 자국경제에 대한 영향 이유

2005 교토의정서 발효
EU, 러시아의 비준으로 교토의정서 발효
- 선진국의 2차 의무(2013-2017) 협상 개시

2009 코펜하겐 총회
개도국, 선진국 모두에 대한 의무 부과 협상 실패 (코펜하겐 총회)
이후 신규 체제도입에 대한 새로운 시작필요

2011 신기후체제 협상
선진국 2차 의무협상과 개도국의무부담 협상 연계로, 2020년 이후에 대한 신기후체제 협상위원회 출범

2012 교토의정서 2차 의무
교토의정서상 선진국 2차의무(2013~2020)타결/일부국가 불참
- 교토의정서 연장(2020까지), GCF 한국 송도 승인- 2020년까지 1천억불 조성

2015 파리협정체결
194개국의 온실가스 감축합의
- 우리나라 또한 온실가스 감축국가로 지정(CO2 배출량 세계 7위)

2020 신기후체제도입
국가별 온실가스 감축목표 달성필요
- 우리나라 2030년 배출전망치 대비 37% 감축

미국 : 2021년 파리협정 재가입

파리 협정

2015 Paris Agreement / 194 Parties / NDC 151 Parties - 40GtCO2eq (2010대비 +5.9%)
선진국과 개도국 구분없이 모든 당사국의 온실가스 감축 참여(자발적)

REDD+ : Reducing Emissions from Deforestation and Forest Degradation, ITMO : International Transferred Mitigation Outcomes (국외 감축실적)
NDC : Nationally Determined Contribution(국가 자발적 기여), LEDS : Long-term greenhouse gas Emission Development (장기 저탄소 발전전략)

파리 협정 (장기목표)

글로벌 온실가스 감축 장기목표를 산업화 이전 대비 2℃ 보다 상당히 낮은 수준 확정
국가별 INDC 제출량과 글로벌 감축목표와의 GAP 해결방안 마련 필요

🔒 Summary
- 지구 평균기온 상승을 산업화 이전 대비 2℃ 보다 상당히 낮은 수준으로 유지하고, 장기적으로 1.5℃로 제한하기 위한 노력 추진
  - 공통의 차별화된 책임원칙에 따라 당사국별 상이한 여건을 고려
  - 당사국은 공동의 장기목표를 달성하기 위해 조속하게 배출량 정점에 도달할 것을 촉구
    단, 개도국의 경우 배출량 정점 도달 시기가 오래 걸릴 수 있음을 인정

↓ Issue

- 2℃ 상승억제를 위해서는 전세계 배출량을 440억tCO₂e로 제한필요

  But 기존 국가별 제출된 INDC를 기반
  2025년에 87억 tCO₂e, 2030년에는 151억 tCO₂e이 더 많을 것으로 예상

- 국가별 INDC의 수정 없이 글로벌 감축목표만을 확정 → 글로벌 감축목표 달성불가
  제5차 기후변화에 관한 정부간협의체(IPCC) 종합보고서 : 2100년 지구온도 상승을 2℃ 이하로 제한하기
  위해 지구가 향후 사용 가능한 탄소예산(Carbon Budget)으로 약 1조톤CO₂e 산출
  → COP26 권고 : 국가별 진전된 NDC의 제출 권고 (우리나라 12월 상향된 NDC 제출완료)

COP : Conference of the Parties (UN기후변화협약 당사국 총회)

파리 협정 (온실가스 감축)

목표제출은 의무화 이나, 목표달성여부는 국가별 자발적 달성 – 국제 법적 구속력없음
COP26 : 5개년단위로 향후 10년간의 온실가스 감축목표 제출필요(2025년 재 제출 ~2035까지)

🔒 Summary
- 목   표 : 최대한 조속히 전 지구적 최대 배출연도 달성
- 의   무 : 모든 국가가 스스로 결정한 감축목표(INDC) 5년 마다 제출
- 구 속 력 : 목표 제출은 의무로 하되, 이행은 각국이 국내적으로 담보
- 진전원칙 : 차기 감축 목표 제출 시, 이전 수준보다 진전된 목표 제시
- 목표방식 : 선진국(경제전반에 걸친 절대적 감축방식), 개도국은 유연한 감축방식 채택

Issue
- 5개년 단위의 자발적 기여방안(INDC) 제출 – 단, 지속적으로 진전된 감축목표로의 제출
- 선진국(절대량방식), 개도국(BAU, 원단위 등) 감축목표 설정방식 인정 (2018년 재평가)
  우리나라 – BAU방식에서 절대량방식으로 감축목표 변경
- 감축목표 달성에 국제 법적 구속력은 없으며, 국가별 국내법에 기반한 목표 달성 추진
  법적 구속력이 없는 상태에서 탄소국경조정제도 등 국가별 제재 조치 시행 예정

파리 협정 (이행 점검)

5개년 단위로 UN주도의 국가별 이행사항을 점검 (감축, 적응, 기술이전 등)
점검결과에 대한 UN 차원의 정보공개, 그러나 이행사항에 따른 패널티 없음

🔒 Summary
- 2023년부터 5년 단위로 파리 협정 이행 및 장기목표 달성 가능성을 평가하는 전 지구적 이행점검(Global Stocktaking) 실시
  ※ COP26 : 보고양식 등 확정
- 종합점검은 개별 국가 단위가 아닌 전지구적 단위의 감축·적응·재정지원 현황 점검 이며, 포괄적이며 촉진적 방식으로 시행 규정
- 이행보고·검토 : 각 국의 온실가스 감축과 지원에 대해 이행 보고하고 점검을 받되, 개도국에게는 보고 범위, 주기, 검토 범위 등 유연성 부여

Issue
- 개도국과 선진국간의 보고내용/방식 등 이견이 존재하였으나 작년 최종확정
  COP26 회의에서 최종적인 보고양식 등 확정 완료

파리 협정 (기술 개발 및 이전)

기술프레임워크 : 온실가스 감축을 위한 기술개발 및 이전의 활동 지침
온실가스 감축을 위한 신기술 R&D 등의 재정 지원 등 가속화 전망

Summary

감축·적응에 기술이 핵심이라는 비전 공유, 기술협력 확대전략 마련 위한 기술 프레임워크 수립

- 장기 비전 : 감축 및 적응대응력 강화를 위한 기술 개발·이전의 중요성에 대해 비전 공유
- 프레임워크 : 기술 개발·이전 촉진을 위한 기술메커니즘 활동에 지침마련 ('기술 프레임워크')
  ※ 실질적 프로젝트를 통한 기술수요평가의 이행 강화 및 재정·기술 지원, 이전가능기술에 관한·평가 등 촉진
- 기술 혁신 : 효과적·장기적 대응에 R&D 협력 및 기술 접근 확대를 기술, 재정 메커니즘을 통해 지원
- 기술메커니즘 : 기후변화대응 기술협력은 기술 메커니즘에 의해 수행
  ※ 메커니즘 강화, 연구개발실증 및 내생적 역량 제고에 추가적 노력
- 협력 강화 : 기술 개발·이전에 관한 협력 강화를 위해 선진국이 재정 지원을 포함하여 지원을 제공
- 역량배양 : 개도국의 효과적인 기후대응 역량 증진을 위해 협력하며, 파리 위원회 설립

Issue

- 선진국은 에너지효율향상, 신재생에너지, 탄소 포집(CCS) 등 온실가스 감축 신기술의 지속적 R&D 산업을 육성하는데 투자할 것으로 전망
- 재정 프레임워크를 연계한 기술의 이전 필요성 : GCF 등을 활용한 기술이전 확대

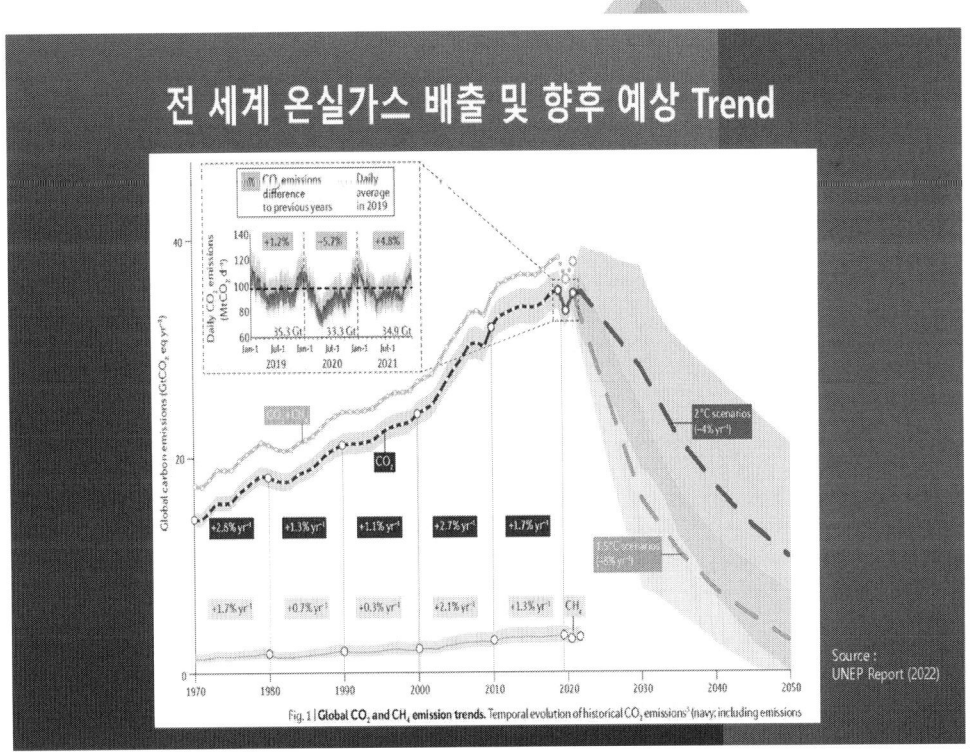

전 세계 온실가스 배출 및 향후 예상 Trend

Source : UNEP Report (2022)

Fig. 1 | Global CO_2 and CH_4 emission trends. Temporal evolution of historical CO_2 emissions (navy: including emissions

10대 경제국 온실가스 배출량 현재와 미래

('21년 COP26 기준)

Source : IPCC 2022 Report (AR6)

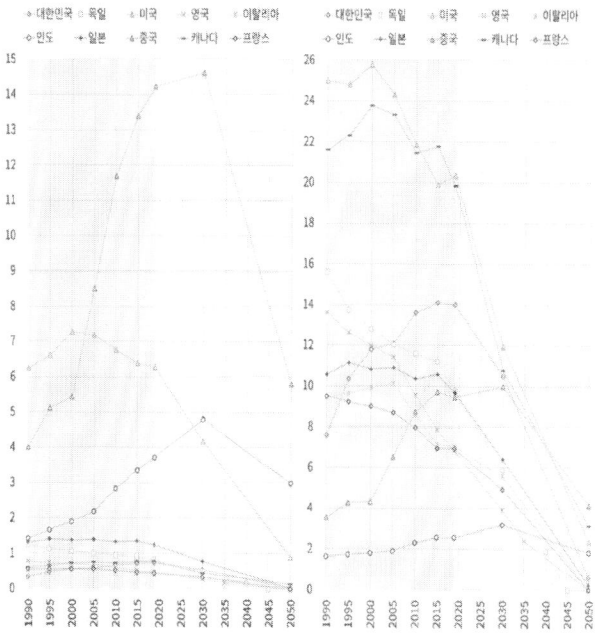

우리나라 온실가스 감축 목표

우리나라는 2016년 이후 3차에 걸쳐 NDC 제출 - 지속적인 온실가스 감축목표 강화
2018년 온실가스 배출량 대비 40% 감축(4,367억톤) ➡ 2.91억톤 감축필요

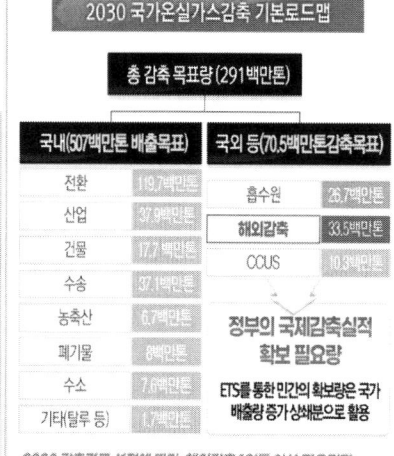

2030 감축경로 설정에 따라 해외감축 1억톤 이상 필요전망

우리나라 온실가스 감축 목표 조정

[2030년 해외 감축목표 207% 상향 조치(2021.10.18)
감축 경로(Pathway)에 따른 필요 해외감축분은 더욱 많을 것으로 예상]

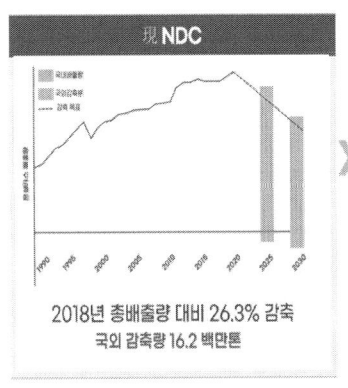

現 NDC

2018년 총배출량 대비 26.3% 감축
국외 감축량 16.2 백만톤

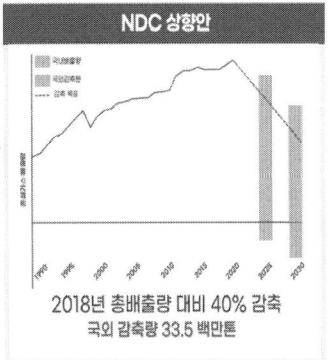

NDC 상향안

2018년 총배출량 대비 40% 감축
국외 감축량 33.5 백만톤

2050년 탄소중립계획과 상향된 NDC에 따른 3차할당계획 조정검토필요

CO2 누적배출량과 지구표면온도 관계

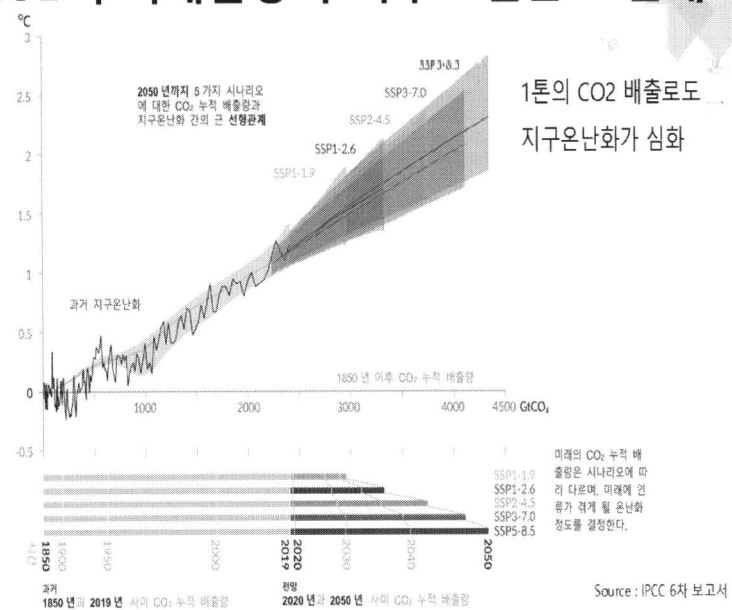

1톤의 CO2 배출로도 지구온난화가 심화

Source : IPCC 6차 보고서

기후변화 Scenario (SSP 기준)

SSP(Shared Socioeconomic Pathways, 공통사회 경제통로) : 2100년 기준 복사 강제력 강도와 함께 미래 기후변화 대비 수준에 따라 인구, 경제, 에너지 사용 등의 미래 사회경제시스템의 변화를 적용한 경로

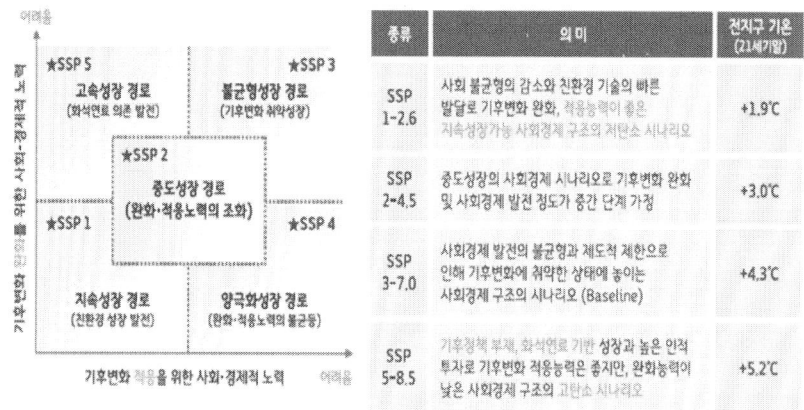

종류	의미	전지구 기온 (21세기말)
SSP 1-2.6	사회 불균형의 감소와 친환경 기술의 빠른 발달로 기후변화 완화, 적응능력이 좋은 지속성장가능 사회경제 구조의 저탄소 시나리오	+1.9℃
SSP 2-4.5	중도성장의 사회경제 시나리오로 기후변화 완화 및 사회경제 발전 정도가 중간 단계 가정	+3.0℃
SSP 3-7.0	사회경제 발전의 불균형과 제도적 제한으로 인해 기후변화에 취약한 상태에 놓이는 사회경제 구조의 시나리오 (Baseline)	+4.3℃
SSP 5-8.5	기후정책 부재, 화석연료 기반 성장과 높은 인적 투자로 기후변화 적응능력은 좋지만, 완화능력이 낮은 사회경제 구조의 고탄소 시나리오	+5.2℃

Source : 기후정보포털 (IPCC 6차 보고서 기반)

기후변화 Scenario (RCP 기준)

RCP(Representative Concentration Pathways, 대표농도경로) :
2100년 기준 복사 강제력에 따른 온실가스 농도 경로

종류	의미	CO_2농도	전지구 기온 (21세기말)
RCP2.6	지금부터 즉시 온실가스 감축 수행	420ppm	+1.3℃
RCP4.5	온실가스 저감정책 상당히 실현	540ppm	+2.4℃
RCP6.0	온실가스 저감정책 어느 정도 실현	670ppm	+2.7℃
RCP8.5	현재 추세대로 온실가스 배출	940ppm	+4.0℃

◉ RCP 숫자는 온실가스로 인한 추가적인 지구흡수에너지량을 의미함

◉ (예) RCP8.5: CO_2 농도가 940ppm이 되면 태양에너지 8.5W/m^2 추가 흡수

Source : 기후정보포털 (IPCC 5차 보고서 기반)

영상자료 https://www.carbonbrief.org

Why 1.5℃ ?

CNN Report Link

기후 변화의 영향
(1950 vs 2010)

Source : IPCC 5차 보고서(2014)

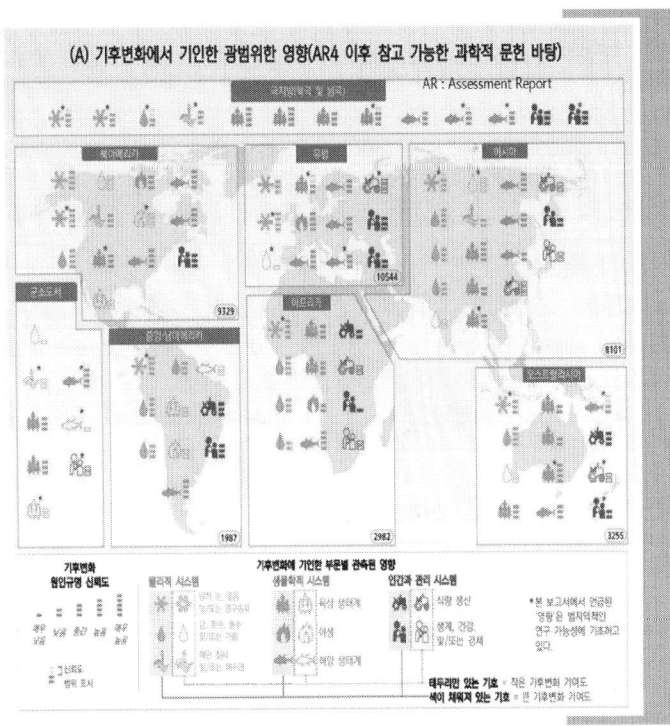

(A) 기후변화에서 기인한 광범위한 영향(AR4 이후 참고 가능한 과학적 문헌 바탕)

기후변화 (지구온난화 수준 모사)

a) 1℃ 지구온난화 수준에서 연평균 온도 변화(℃)

1℃ 지구온난화는 모든 대륙에 영향을 미치며 관측 자료와 모열 결과 모두 대체로 해양보다 육지에서 온난화가 뚜렷하게 나타난다. 대부분의 지역에서 관측과 모델 결과가 일치한다.

b) 1850~1900년 대비 연평균 온도 변화(℃)

모든 지구온난화 수준에서 육지가 해양보다 더 많이, 북극과 남극이 열대보다 더 많이 온난화된다.

Source : IPCC 6차 보고서

기후변화 (강수량 및 토양수분함량 변화 모사)

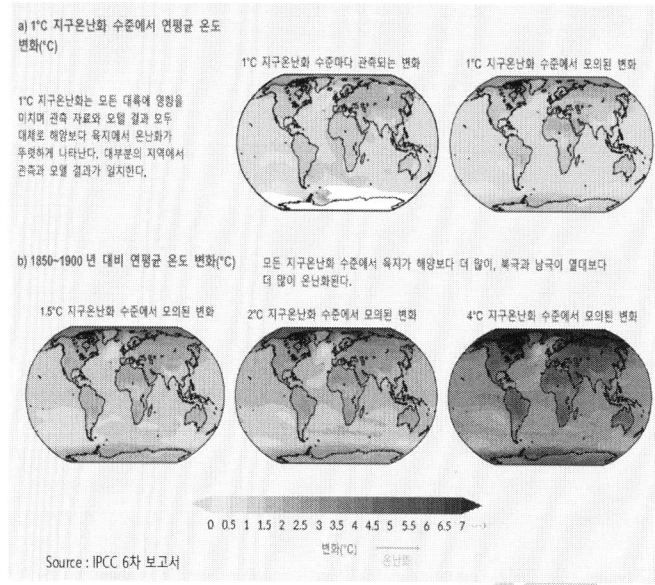

c) 1850~1900년 대비 연평균 강수량 변화(%)

강수량이 고위도 지역과 적도 부근 태평양, 일부 온순 지역에서는 증가하지만 아열대 지역과 열대 지역에서는 부분적으로 감소한다.

상대적으로 작은 절대 변화가 건조한 지역에서는 큰 비율(%)의 변화로 나타날 수 있다.

d) 연평균 전층 토양수분함량 변화(표준편차)

모든 지구온난화 수준에서 토양수분함량 변화는 대체로 강수량 변화에 뒤이어 발생하지만 증발산량의 영향에 의해 차이가 나타나기도 한다.

상대적으로 작은 절대 변화도 기후조건의 연간 변동성이 낮은 건조 지역에서 표준편차 단위로 표현되면 커 보일 수 있다.

Source : IPCC 6차 보고서

지구온난화에 따른 극한현상 빈도와 강도 (육지 온도)

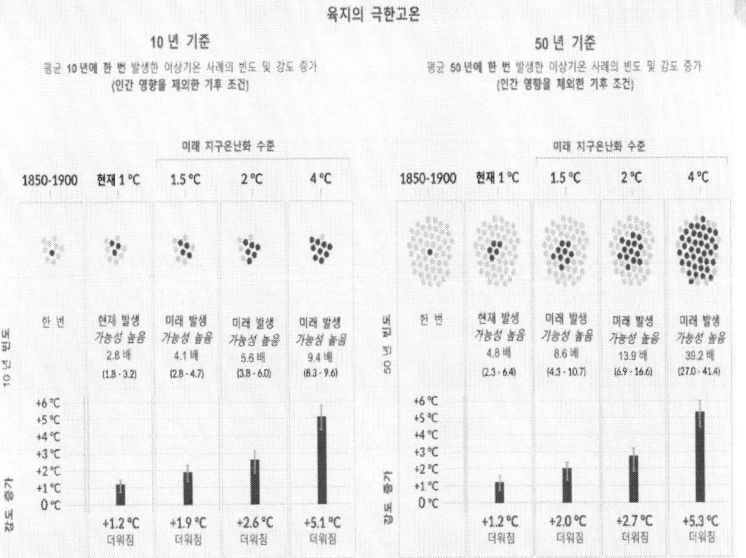

Source : IPCC 6차 보고서

지구온난화에 따른 극한현상 빈도와 강도 (육지 호우 및 가뭄)

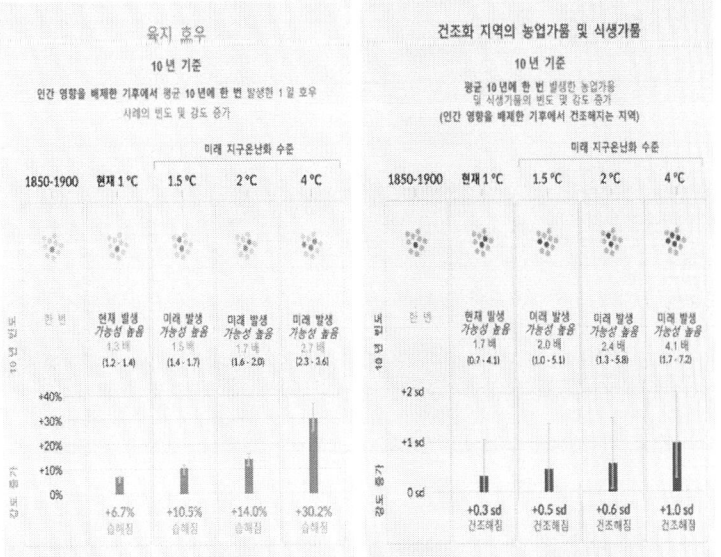

Source : IPCC 6차 보고서

참고 온실가스 1톤의 의미

- 가정집 연간 전기사용에 따른 온실가스 배출량

  약 2톤(1.96톤)

  ※ 한달 350kWh 사용기준

- 자동차 1년 주행에 따른 온실가스 배출량

  약 4.2톤(4.21톤)

  ※ 20,000km, 휘발유, 연비 10km/L 기준
  ※ 서울-부산 왕복 시 0.17톤 배출

- 소나무 1그루 연간 온실가스(CO_2) 흡수량

  약 6.6kg

  ※ 30년생 소나무 기준
  ※ 1톤을 흡수하기 위해서는 약 150그루 필요

- 우리나라에서 가장 많은 온실가스를 배출하는 기업

  약 8,000만톤

  ※ 2019년 배출량 기준
  ※ 4,000만가구의 연간 전력사용량 수준

- 참이슬 소주 한병의 생애주기(LCA) 배출량

  약 0.2 kg

  ※ 참이슬 오리지널 360ml 기준

- 태양광 발전 1MW 전력생산에 따른 연간 온실가스 감축량

  약 612톤

  ※ 1MW 설비 면적 = 10,000㎡
  ※ 축구장면적 : 약 7,140㎡

- 서울-부산, 온실가스가 가장 적게 배출 되는 교통수단은?

  철도(5.9kg/인)

  ※ 항공 53.3, 버스 10, 승용차(휘발유) 50, 승용차(하이브리드) 38

- 생애주기 배출량이 가장 적은 생수 (500mL, PET)는?

  크리스탈 생수 (0.074kg/개)

  ※ e블루 0.08, 평창수 0.1, 삼다수 0.12

온실가스 1톤의 의미를 알면, 우리 생활에서의 온실가스를 줄일 수 있습니다.

- 출처 : 환경성적표지 유효 인증현황('21.5월 기준, 한국환경산업기술원)

탄소 중립

탄소중립의 범위

탄소중립은 "자발적"이며, 탄소중립의 대상 및 범위가 별도로 정해져있지 않습니다.

전 세계 탄소 중립 관련 현황

Net zero emissions race

Source : Earth.org (2021)

EU 탄소중립 6대 주축분야

분야	주요 내용
에너지	- 재생에너지 사용 확대 - 에너지 효율 제고를 위한 에너지법 개정 ('21.06) - 에너지 및 기후변화 계획 개선(~'23)
산업	- 순환경제로의 전환 - 재활용 시장 확대 - 탄소배출 정보 공유
건물	- 건물 Renovation을 통한 에너지 효율 개선
수송	- 스마트 수송전략 (저탄소 차량 전환 및 충전시설 확대) - 해양/항공의 탄소배출거래제 편입
농-식품	- 농장에서 식탁까지의 전략 - 친환경 기술-생산-소비 체계의 도입 및 촉진 - 식품분야 환경위험평가 및 친환경 농식품 소비 촉진
생물다양성	- 생물다양성 전략 및 실행계획 제안 ('21) - 삼림 및 어업분야의 자연훼손 방지책 마련 - 삼림 조성 및 복원

미국 기후변화 게임 체인저 10대 기술영역

분야	주요 내용
건물	- 탄소중립 건설재료 사용 및 탄소중립 비용 달성 건물
전력저장	- 기존 기술 대비 1/10 가격 수준의 에너지저장시스템 개발(ESS)
전력망	- 고도화된 최첨단 에너지시스템 관리기술 (전력망 계획 및 운영)
수송 1	- 저비용, 저탄소 차량 및 교통시스템
수송 2	- 항공, 선박용 저탄소 연료 및 효율 향상 기술
냉난방	- 온실가스 영향이 없는 냉매 개발 및 냉방, 히트펌프 시스템
산업	- 철강, 콘크리트, 석유화학 공정 저탄소화 또는 탄소 Zero
수소	- 그린수소 생산 (탄소배출 Zero 수소 생산)
토양	- 토양, 식물학, 농업기술을 활용한 대기 중 CO_2 제거 및 토양 저장 기술
CCUS	- CO_2 직접 포집 및 활용-저장 기술 개발

CCUS : Carbon Capture, Utilization and Storage

일본 그린성장전략 14대 분야

구분	분야	주요 내용
전력-에너지	해상풍력	- '40년까지 생산능력 4,500만 Kw(원전 45기분) 확대
	암모니아	- '30년까지 암모니아 화력발전 20%로 확대
	수소	- '50년까지 수소 소비량 2,000만 톤으로 확대
	원자력	- 신형 원자로 기술개발 및 국제협력 강화
수송-제조	자동차	- '35년까지 모든 신차 전동화 ('30년까지 이차전지 1만엔/KWh 이하 달성)
	반도체	- '30년까지 파워반도체 소비전력 50% 감축
	물류	- 항만 등의 탈탄소화 추진
	농림수산	- '50년까지 농림수산업 이산화탄소 배출 Zero
	항공	- 전동화 및 대체연료 기술 개발
	카본 재활용	- 효율성 증대 및 비용 절감
가정-오피스	주택	- '30년까지 신축 주택 이산화탄소 배출량 평균 Zero
	자원순환	- Biomass (사탕수수, Palm, 해조류 등의 연료화) 활용 확대
	Life Recycle	- 지역별 탈탄소 비즈니스 추진

우리나라 탄소중립 목표

- 2030년까지 2018년(727.6백만톤) 대비 40% 감축(291백만톤), 2050년까지 탄소중립
  - NDC 및 LEDS를 통해 국제사회에 우리나라 온실가스 감축목표 및 탄소중립 목표 선언
  - (`21.10.18) 2050년까지 국내 순배출량을 '0'으로 하는 2개 시나리오 발표

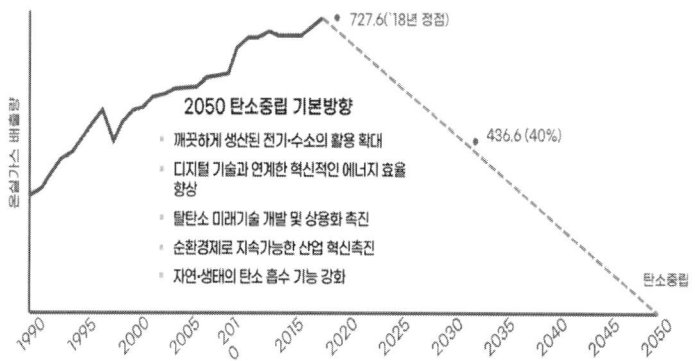

NDC : Nationally Determined Contribution(국가 자발적 기여), LEDS : Long-term greenhouse gas Emission Development Strategy (장기 저탄소 발전전략)

우리나라 탄소중립 Scenario

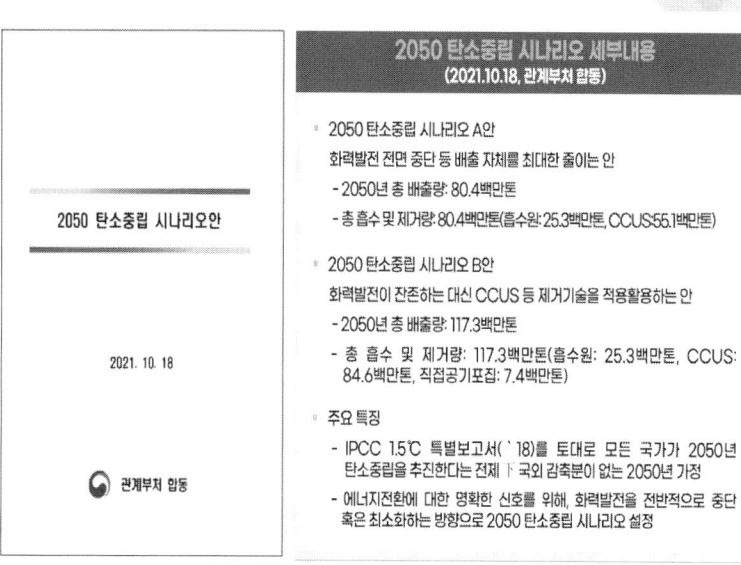

탄소중립 10대 핵심기술 개발전략

Source : 탄소중립 기술혁신 추진전략, 에너지기술원

탄소중립 10대 핵심기술 개발전략

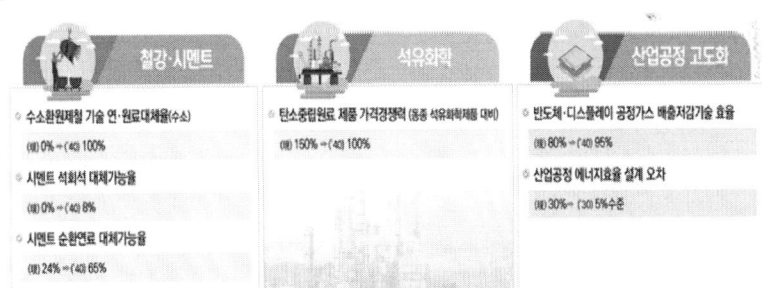

에너지 전환 부문 : 석탄발전 중단, 재생에너지 확대, 수소기반 발전
산업 부문 : 철강 / 석유화학-정유 / 시멘트 산업의 에너지 대체 및 효율화
수소 부문 : 수소 생산 탄소 Zero 기술 개발, 수소 생산 단가 저감
CCUS 부문 : 이산화탄소 포집-저장-활용 기술 개발

Source : 탄소중립 기술혁신 추진전략, 에너지기술원

우리나라의 탄소 감축정책 방향

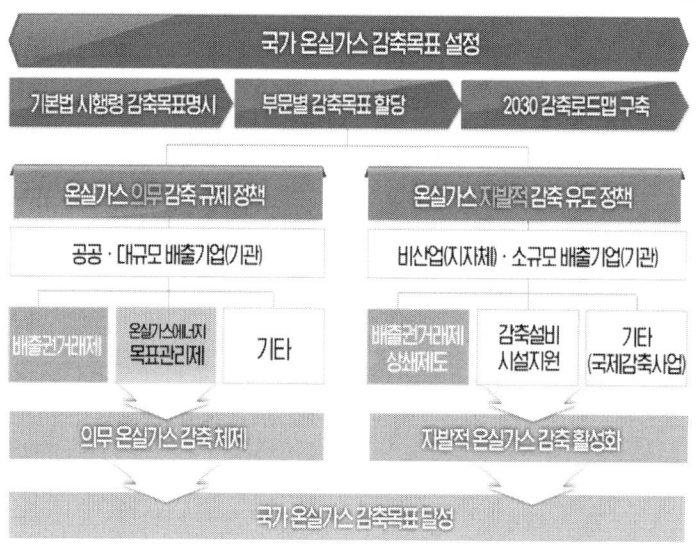

EU 주관 탄소국경조정제도

[탄소국경조정제도(CBAM) 란
carbon leakage* 문제 해결을 위해 수입품에 비용을 부과하는 제도]

* carbon leakage: 탄소배출량 감축 규제가 강한 국가에서 상대적으로 규제가 약한 국가로 시설, 투자 등을 이전하는 것을 의미

- (목적) 환경규제가 느슨한 수입 제품을 제한함으로써 국가별 공정한 경쟁을 유도
- (인증서) 연단위, 수입업자가 제출
- 이산화탄소 배출량 1TON 당 CBAM 인증서 1개
- (인증서 가격) EU-ETS의 배출권 가격*과 CBAM 인증서 가격 연동
   * EU ETS 배출권 경매 종가의 주당 평균 가격을 적용
- (인증서 제출) EU로 수출하는 수출품의 전년도 탄소 배출량과 이에 상응하는 CBAM 인증서 수를 EU에 제출

[5대 부문(철강, 시멘트, 알루미늄, 비료, 전기)을 적용
`23.1.1. CBAM 시범 도입, `26.1.1. CBAM 본격 시행]

CBAM : Carbon Border Adjustment Mechanism
ETS : Emissions Trading System

탄소국경조정제도에 따른 국내 산업 영향

> EU 탄소국경조정제도 대상품목
> **철강부문 수입국 5위, 한국**

〈EU 탄소국경조정제도 대상품폭 對EU 수출 현황〉

(단위: 백만불, 톤)

품목	2018 금액	2018 물량	2019 금액	2019 물량	2020 금액	2020 물량
철·철강	2,485	2,946,121	2,124	2,763,801	1,523	2,213,680
알루미늄	110	30,652	155	46,892	186	52,658
비료	1	967	1	8,005	2	9,214
시멘트	0	73	0	24	0	80
전기	0	0	0	0	0	0

출처: 산업부 보도자료(2021.7.15.자)

- EU의 수출이 많은 철강업계의 피해가 예상됨
  - 전세계 철강 제품 수출이 11.7% 감소, 우리나라 철강 제품 생산은 0.25% 감소 추정 (KIEP, 2021)
  - (예시) EU가 1톤당 30유로 과세할 경우, 국내 기업 1.9% 관세율 적용과 유사한 비용

- 피해 산업 업종은 2026년부터 CBAM이 본격 시행 할 경우, 전체 업종으로 확대될 가능성이 존재

유럽이 주도하는 탄소세 정책

탄소세 정의

- (배경) 지구온난화 방지를 위한 온실가스 감축 정책 필요성 확대되어 유럽국가 중심으로 온실가스를 감축하기 위한 수단으로 도입
- (정의) 환경세의 일종으로 탄소를 연소하여 온실가스를 배출하는 일체의 과정에 일정 세율을 부과하는 제도
- 넓은 범위에서 CO_2를 배출하는 화석연료에 대하여 탄소 함유량에 비례하여 부과되는 물품세와 실제 CO_2 배출량에 따라 부과되는 탄소배출세를 포함
- 생산단계 CO_2를 배출하는 기업에 주로 부과되어 간접세 형태로 운영되며, 일부 국가는 소비단계에서 탄소세를 부과
- 정책 등의 영향으로 세금의 규모가 결정

● 직접세 vs 간접세

구분	직접세	간접세
정의	납세의무자와 담세자가 일치	납세의무자와 담세자가 불일치
예시	소득세, 법인세, 상속세, 증여세	부가가치세, 특별소비세 등
장점	소득재분배 기능 ↑	조세저항 ↓ 조세의 행정·순응 비용 ↓ 정부의 조세 확보 비교적 용이
단점	조세저항 ↑	누진세율 적용 어려움 저소득자의 소득대비 세부담 ↑

〈탄소세 직접세/간접세 예시〉
- 호주 : CO_2 다배출 기업에 탄소세를 적용하여 간접세 형태로 운영
- 프랑스 : 화석연료를 소비하는 모든 가정과 기업에 적용하여 직접세 형태로 운영

〈납세의무자와 담세자〉
- 납세의무자 : 세금을 납부할 의무가 있는 사람
- 담세자 : 세금을 최종적으로 부담하는 사람

탄소세 사례

탄소세 적용 국가들

- (참여국) 27개 나라 탄소세를 시행중이며, 미국 북서부 지역과 캐나다 일부 州는 별도로 운영 중
- (탄소세율) 폴란드 0.1US$/tCO2e로 최저, 스웨덴 137.2US$/tCO2e 최대 적용

● 국외 탄소세율

연번	국가또는관할지역	탄소세율 (USD/tCO2e)	2020년수입 (million USD/년)	배출량 커버리지	연번	국가또는관할지역	탄소세율 (USD/tCO2e)	2020년수입 (million USD/년)	배출량 커버리지
1	아르헨티나	5.5	1	20%	17	노르웨이	3.9-69.3	1,758	66%
2	캐나다	31.8	3,407	22%	18	폴란드	0.1	6	4%
3	칠레	5	165	39%	19	포르투갈	28.2	276	29%
4	콜롬비아	5	29	24%	20	싱가폴	3.7	144	80%
5	덴마크	23.6-26.1	576	35%	21	슬로베니아	20.3	147	50%
6	에스토니아	2.3	2	6%	22	남아프리카	9.2	43	80%
7	핀란드	62.3-72.8	1,525	36%	23	스페인	17.6	129	3%
8	프랑스	52.4	9,632	35%	24	스웨덴	137.2	2,284	40%
9	아이슬란드	19.8-34.8	53	55%	25	스위스	101.5	1,239	33%
10	아일랜드	39.3	580	49%	26	영국	24.8	948	23%
11	일본	2.6	2,365	75%	27	우크라이나	0.4	31	71%
12	라트비아	14.1	5	3%	28	캐나다(브리티시컬럼비아)	36.8	1,266	78%
13	리히텐슈타인	101.5	6	26%	29	캐나다(뉴브런즈윅)	31.8	99	39%
14	룩셈부르크	23.5-40.7	N/A	65%	30	캐나다(뉴펀들랜드와래브라도)	23.9	46	47%
15	멕시코	0.4-3.2	230	23%	31	미국의 북서부 지역	23.9	15	79%
16	네덜란드	35.2	N/A	12%	32	캐나다의 프린스 에드워드 아일랜드	23.9	10	55%

기후변화 관련 국제 프로그램

RE100 개요

❖ 기업이 필요로 하는 전력의 100%를 태양광, 풍력 등 친환경 재생에너지로 사용하겠다고 선언하는 자발적인 글로벌 기업 리더십 이니셔티브

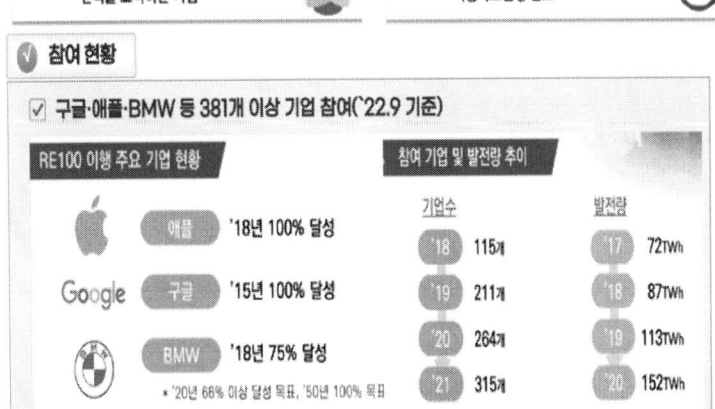

RE100 참여 현황

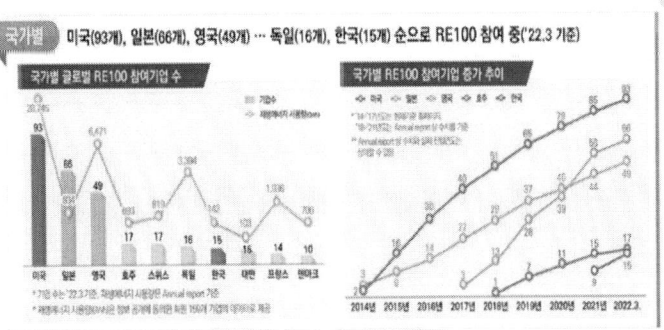

* 2022년 9월 기준, 삼성전자를 포함하여 23개 국내 기업이 글로벌 RE100에 가입

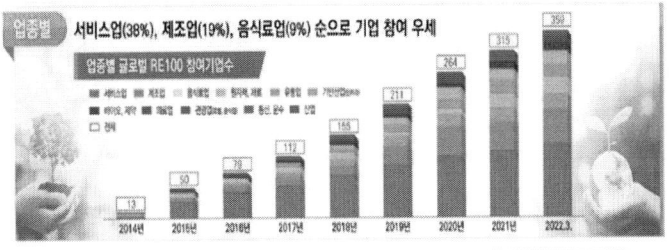

글로벌기업 자발적 배출권 활용사례

Microsoft

● **탄소감축 목표 및 전략**
- 목표: 2030년까지 탄소중립 초과(Carbon Negative) 달성
- 배출되는 탄소보다 더 많은 양을 제거해 순 배출량을 마이너스로 만드는 정책
- 2050년까지는 회사가 설립된 해인 1975년 이후 배출한 모든 탄소를 제거하는 것을 목표로 함

● **Microsoft's pathway to carbon negative by 2030**

● **탄소감축 정책**
- 기업 운영 과정에서 발생하는 탄소 배출량을 2030년까지 절반으로 감축
- 전 세계적으로 상쇄 프로젝트를 통해 탄소 추가 제거
- 2020년 탄소 배출량을 6% 감축
- 탄소 제거 백서를 제시하고 그에 따른 26개 프로젝트 공개
  · 산림 조림/재조림, CCUS 사업 등

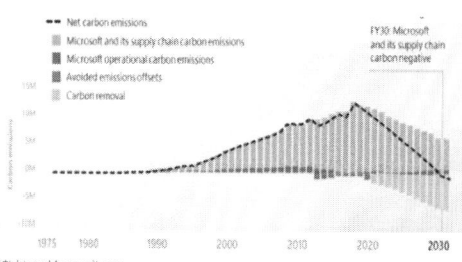

출처: blogs.Microsoft.com

글로벌기업 자발적 배출권 활용사례

● **탄소감축 목표 및 전략**
- 목표: 2030년까지 탄소중립 달성 (Scope 3까지)
- 밸류체인 및 제품 생애주기 전체 배출량의 75%에 대해 자체 감축을 목표로 함
- 잔여 25%의 배출량에 대해 Restore Fund를 통해 상쇄할 예정

● **탄소감축 실천현황**
- 2020년 4월부터 Scope 1+2 에 대해 탄소 중립을 달성함

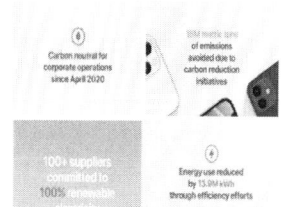

● **Restore Fund**

Conservation International 와의 제휴를 통해 전 세계 삼림 및 자연 생태계의 복원과 보호에 투자하는 $200M 규모의 탄소 솔루션 펀드를 출범시킴

- 펀드의 운용은 Goldmansachs에서 맡으며, Apple과 Conservation International에서 대부분의 금액을 출자하지만, 일부 다른 투자자들도 모집 예정
- 주로 조림 프로젝트를 중심으로 NCS 프로젝트에 투자할 예정이며 주로 Verra 인증을 받은 프로젝트로 포트폴리오를 구성할 예정임
- 초기 단계에는 연간 최소 1M tCO2e 감축을 목표로 펀드를 운용할 예정

출처: 애플 뉴스룸 애플 Environmental Progress Report 2021

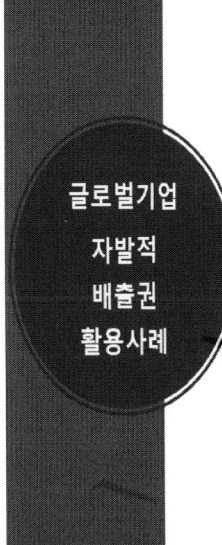

글로벌기업 자발적 배출권 활용사례

● 탄소감축 목표 및 전략
- 목표: 2050년까지 탄소중립 달성
  (25년까지 '15년 대비 30% 감축)
- Reduce, Convert, Compensate의 3단계 전략을 통해 탄소감축 및 상쇄의 원칙과 일치하는 감축 방향성을 가짐
- 사용단계에 많은 온실가스가 배출되는 산업의 특성에 걸맞게 자동차 1대의 생애주기를 단위로 탄소집약도를 집계 및 관리함

● 탄소감축 실천현황
- 2021 온실가스 배출량: 9,593,113 tCO2e (scope 1+2)
- 지속적으로 자동차 1대의 생애주기 단위 탄소집약도를 낮추고 있음

● Volkswagen의 탄소배출량 상쇄 원칙
- 전기차로의 상품 포트폴리오 전환 및 생산공정 개선을 통해 배출량의 자체 감축을 최우선으로 삼고 있으나, 불가피한 배출량에 대해 **고품질의 자발적 상쇄배출권**을 통해 상쇄하는 것을 원칙으로 함
- 현재 세계에서 가장 큰 규모의 조림 및 생태계 보존 프로젝트인 Kantingan Mentaya Project에 투자 중임
- 차기 상쇄 프로젝트로 VCS, GS 인증을 받은 환경 및 지역사회 개선 효과가 큰 Agriculture, Forestry, Renewable Energy 프로젝트에 투자를 계획 중임
- 자발적 배출권 사업 구매 및 투자 포트폴리오 구성

국가	인증기관	사업 내용	2021 획득량 (tCo2e)
산림보존 케냐	Verra	Kasigau Corridor 지역 REDD 프로젝트	19,437
풍력 인도	GS	Tamil Nadu 지역 200MW 풍력발전 프로젝트	350,633
산림보존 인도네시아	Verra	Kantingan Mentaya 지역 산림복원 및 보존 프로젝트	865,081

출처: 폭스바겐 홈페이지, CDP Public Disclosure, Volkswagen Climate Change 2021

SAMSUNG

● 탄소감축 목표 및 전략
- 목표: 2050년까지 탄소중립 달성
- 제품의 Life-Cycle에 따른 탄소 감축 방안 구축

사업장 온실가스 감축	반도체 생산공정의 F-Gas 감축설비 운영 / 제조 공장 에너지 효율 개선 프로젝트 실행 / 재생에너지 사용 확대
제품 사용단계 온실가스 감축	에너지 고효율 제품 개발로 사용단계의 온실가스 배출량 감축 / 이전 세대 제품대비 전력 효율이 10% 이상 개선된 저전력 반도체 개발
기타 밸류체인 온실가스 감축	협력회사, 물류, 임직원 출장, 전기차 확대 등의 온실가스 배출량 관리
사업장 외부 온실가스 감축	CDM 등 외부 감축 프로젝트를 통해 탄소배출권 확보

● 탄소감축 실천현황
- 2020 온실가스 배출량: 14,806,000 tCO2e (scope 1+2)
- 2020년 미국, 유럽, 중국 사업장 100% 재생에너지 사용
- Scope 3 온실가스 감축량: 3억 1백만 tCO2e (~'20년 누적)

출처: 삼성전자 뉴스룸, 삼성전자 지속가능보고서 2021

● Go Eco Initiative

Carbon Footprint Ltd 와 협업을 통해 가전제품 사용에 따른 Scope 3 온실가스 배출량을 상쇄하기 위한 삼성전자의 글로벌 이니셔티브

- 해당 이니셔티브에 포함된 모든 상쇄사업은 Verra를 통해 인증받은 자발적 배출권 프로젝트로 구성됨
- 2021년도에 시작된 프로젝트로, 작년 한 해 동안 브라질과 인도에서 270,948 tCO2e를 상쇄하였으며, 영국과 아일랜드에서 8,126그루의 나무를 심음
- 상품 포트폴리오 구성

종류	국가	인증기관	사업 내용
태양광	인도	Verra	Andhra Pradesh 지역 500MW 태양광 패널 설치 사업
풍력	인도	Verra	Tamilnadu 지역 396개 터빈에 해당하는 집단 풍력 에너지 사업
산림보존	브라질	Verra	Portel Para 지역 아마존 열대우림 보존 사업
조림	영국	Verra	조림을 통한 지역사회 토종 생태계 보존 사업

Greenwashing Issue

[2021년 8월
Shell의 탄소중립 제품의 캠페인 방식이 그린워싱이라고 제소

Shell campaign promoting carbon offsetting is greenwashing, Dutch advertising watchdog rules

Shell의 Carbon Neutral (탄소중립) 제품 소송

네덜란드 소재 대학교 9명의 대학생들이 Shell의 탄소중립 제품 캠페인(Drive Co2 Neutral)이 그린워싱이라며 네덜란드 광고위원회에 제소

- Shell이 고객들에게 추가적 비용을 내고 선택하게 한 탄소중립 제품 상쇄크레딧에 관한 의문 제시
- Shell이 제공하는 크레딧 가격이 EUR 0.01/litre 인데, 과연 한 석연료 사용에 따른 온실가스 배출을 상쇄할 수 있는 것인가에 대한 해명 요구
- 실제로는 상쇄 크레딧 중에 고품질에 해당하는 조림 크레딧도 있는 것으로 확인 됨

Greenwashing은 기업이나 단체에서
실제로는 환경보호 효과가 없거나
심지어 **환경에 악영향을 끼치는 제품을 생산**하면서도
허위·과장 광고나 선전, 홍보수단 등을 이용해
친환경적인 모습으로 포장하는 '위장환경주의' 또는 '친환경 위장술'

Greenwashing Issue

[2020년 11월
Shell이 탄소중립에 사용하는 REDD 크레딧에 대해 문제 제기

"Worse than doing nothing": Shell's REDD offsets in Indonesia and Peru

Shell의 REDD Offset 문제 제기

Shell이 덴마크 운전자들에게 탄소중립 휘발유를 구매할 수 있도록 주유소에서 판매를 시작하였음. 3명의 덴마크 저널리스트들이 확인한 결과 Shell의 탄소중립 휘발유를 믿을 수 없다는 결론을 내림, 해당 크레딧을 상쇄사업을 "안한 것만 못하다" 는 결론.

- 제거(Removal) 크레딧이 아닌 회피(Avoidance) 크레딧을 사용하여 과연 상쇄가 되는 것인가?
- 인도네시아와 페루에서 회피된 탄소가 덴마크 탄소 배출에 상쇄되어 탄소중립이 되는 것이 과연 가능한 것인가?
- 탄소누출(Leakage) 추가성(Additionality) 등에 대한 이슈들이 있다고 설명

Greenwashing Issue

[2022년 10월
SK 엔무브의 탄소중립 윤활유과 과대광고로 광고위 제소]

Korean Oil Firm Faces Greenwashing Claims Over Carbon-Offset Ads

- SK's Uruguay project criticized for limited climate benefits
- Global companies come under scrutiny for making misleading ads

Activist groups in South Korea have accused SK Lubricants Co. of greenwashing, alleging the company is using an unreliable carbon-offsetting project to advertise its products.

In an ad campaign in September, SK said customers can buy carbon-neutral engine lubricants because it's offsetting emissions with high-quality carbon credits from a reforestation project in Uruguay. Solutions for Our Climate and Consumers Korea are bringing separate claims to regulators, saying the project has been criticized for offering no environmental benefits.

SK 엔무브의 Carbon Neutral (탄소중립) 제품 과대광고 제소

한국의 환경단체에서 SK엔무브의 탄소중립 윤활유가 그린워싱이며 과대광고라며 광고위에 제소

- 해당 윤활유가 사용됨으로 발생되는 온실가스에 대한 정확한 경계 설정 및 산출방식 공개 미비
- 해당 윤활유가 탄소중립에 달성되기 위해 소각된 크레딧 양에 대한 설명 누락
- 사용된 크레딧이 고품질 조림 크레딧이라고 밝혔으나, 최근 한 Rating 기관에 의해 추가성이 없어 해당 크레딧 품질이 최저인 0를 받은 것으로 확인 됨

2023 CCPI
Climate Change Performance Index

RESULTS

Monitoring Climate Mitigation Efforts of 59 Countries plus the EU – covering 92% of the Global Greenhouse Gas Emissions

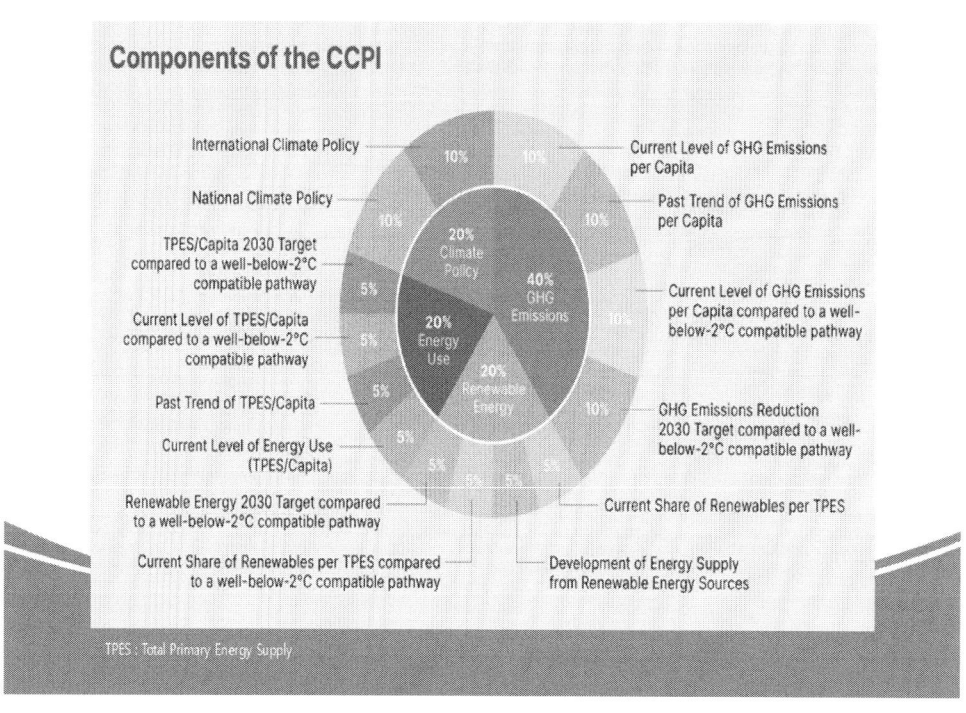

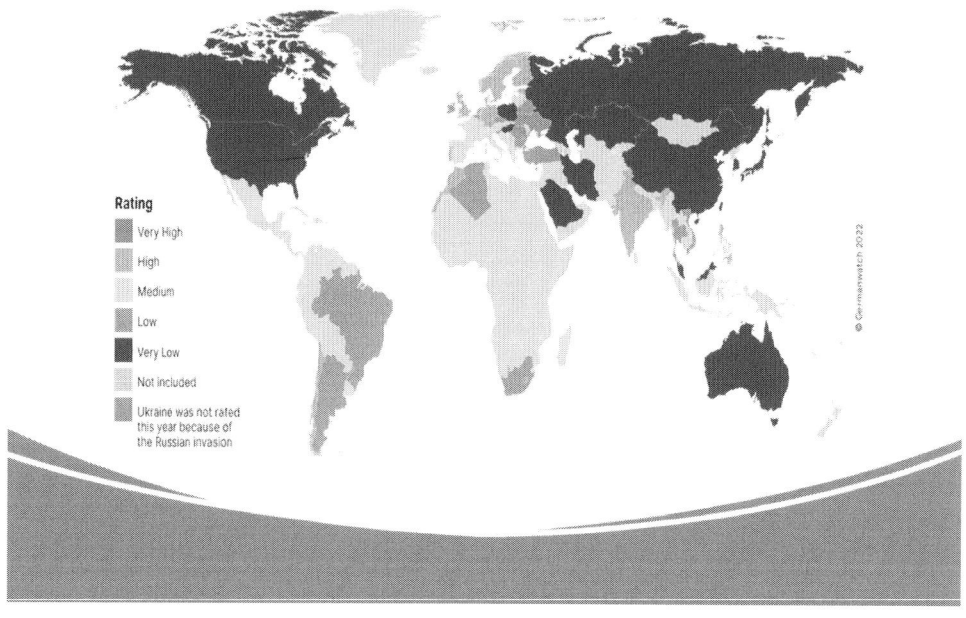

Overall Results

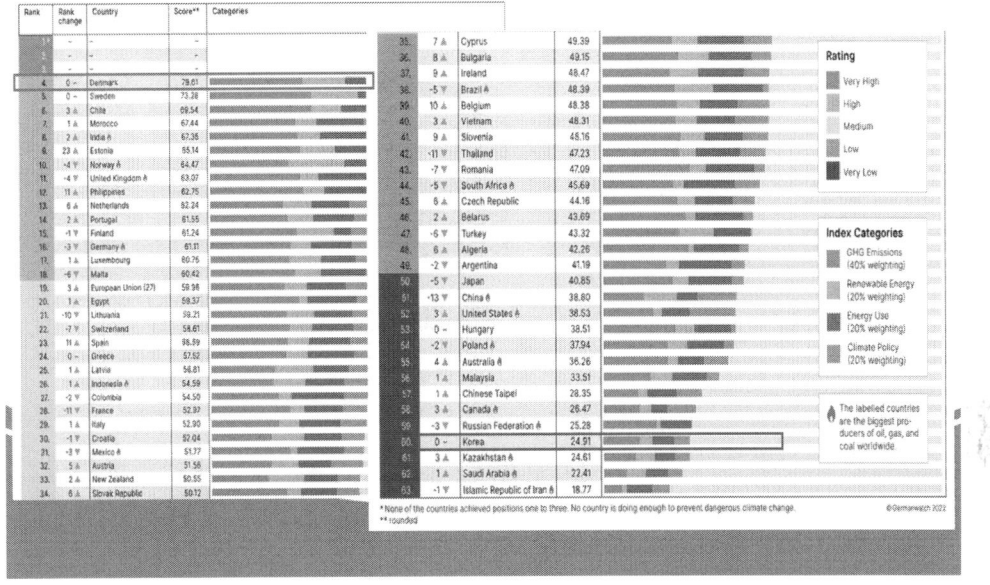

GHG Emissions

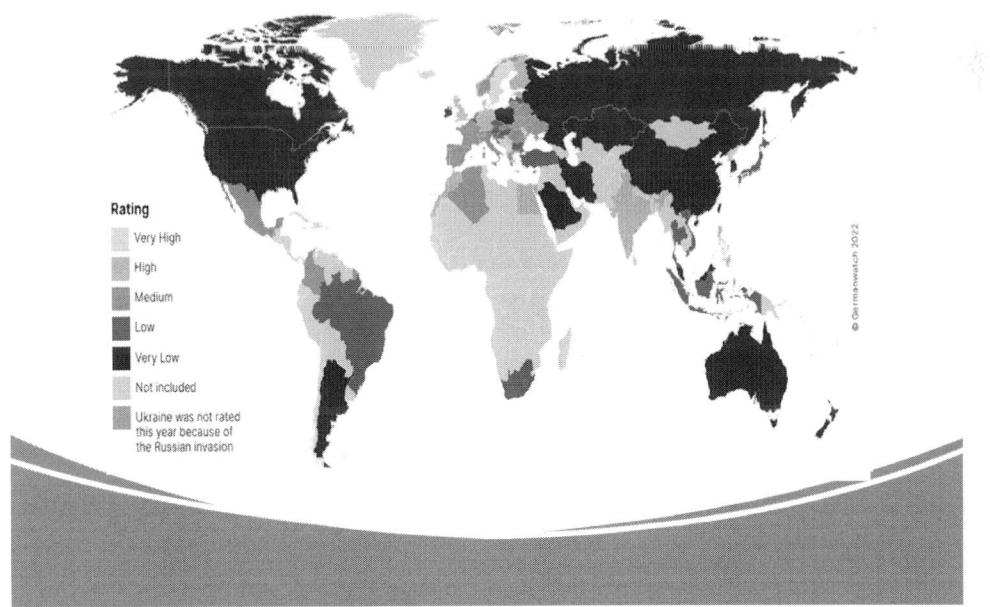

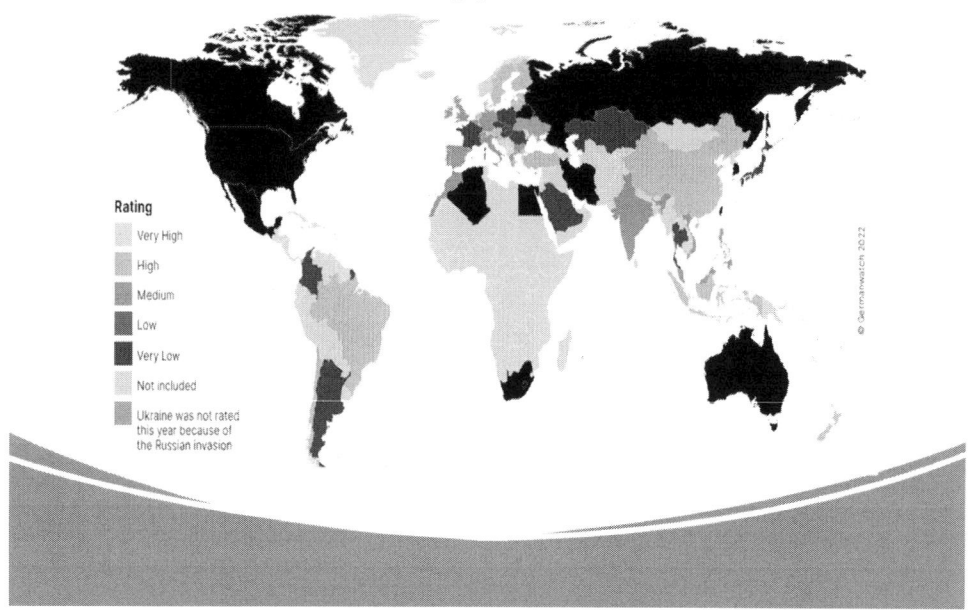

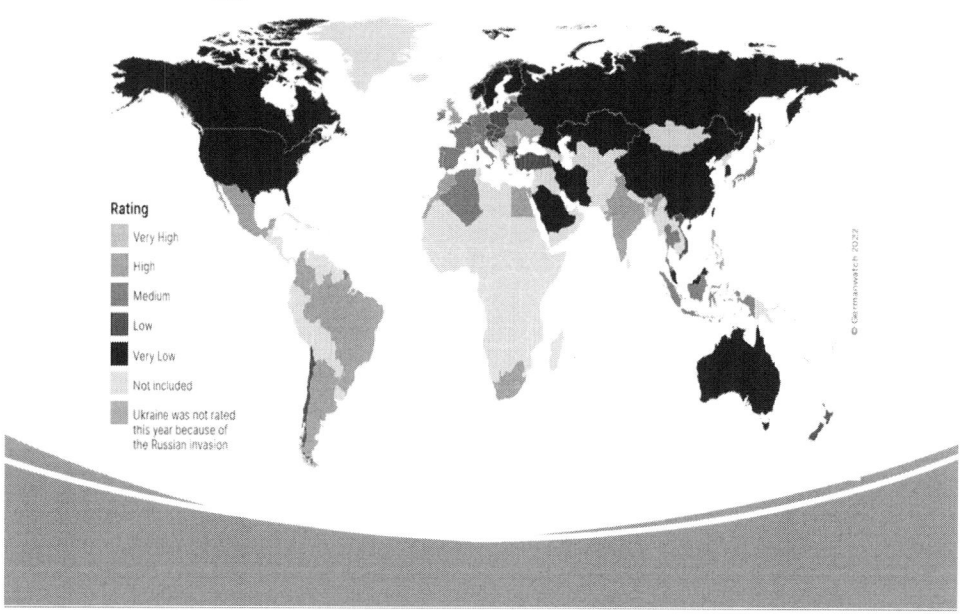

Climate Policy

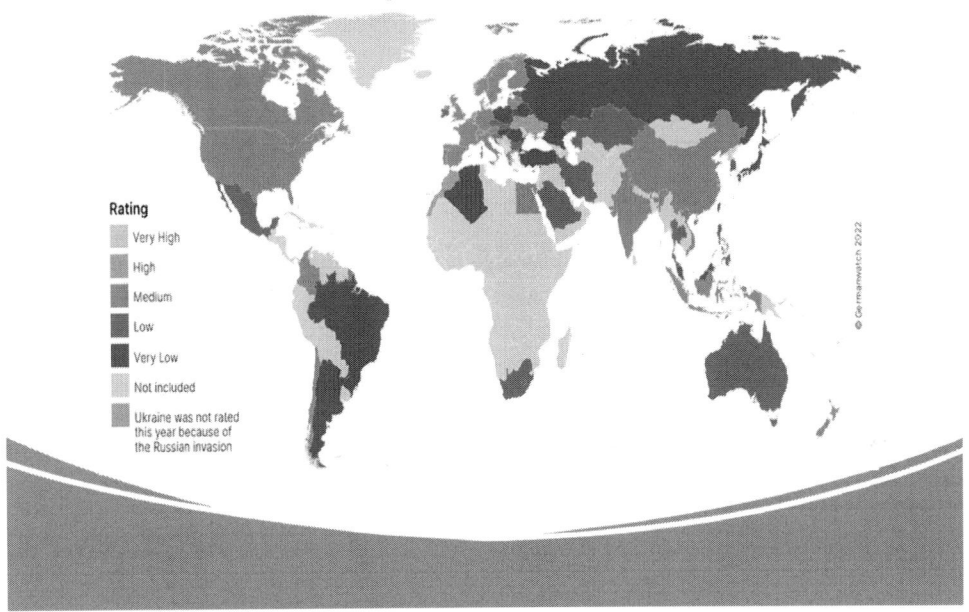

Rating Tables in categories

Korea (Ranking/Index score)

Greenhouse Gas	: 56th (10.51)	Chile	: 2nd (34.50)
Renewable Energy	: 53rd (3.49)	Norway	: 3rd (19.35)
Energy Use	: 60th (5.93)	Colombia	: 4th (17.71)
Climate Policy	: 50th (4.98)	Denmark	: 4th (20.00)
Overall Score	: 60th (24.91)	Denmark	: 4th (79.61)

참고 온실가스 1톤의 의미

- 가정집 연간 전기사용에 따른 온실가스 배출량
  약 2톤 (1.96톤)
  ※ 한달 350kWh 사용기준

- 자동차 1년 주행에 따른 온실가스 배출량
  약 4.2톤 (4.21톤)
  ※ 20,000km, 휘발유, 연비 10km/L 기준
  ※ 서울-부산 왕복 시 0.17톤 배출

- 소나무 1그루 연간 온실가스 (CO_2) 흡수량
  약 6.6kg
  ※ 30년생 소나무 기준
  ※ 1톤을 흡수하기 위해서는 약 150그루 필요

- 우리나라에서 가장 많은 온실가스를 배출하는 기업
  약 8,000만톤
  ※ 2019년 배출량 기준
  ※ 4,000만가구의 연간 전력사용량 수준

- 참이슬 소주 한병 의 생애주기(LCA) 배출량
  약 0.2 kg
  ※ 참이슬 오리지널 360ml 기준

- 태양광 발전 1MW 전력생산에 따른 연간 온실가스 감축량
  약 612톤
  ※ 1MW 설비 면적 = 10,000㎡
  ※ 축구장면적 : 약 7,140㎡

- 서울-부산, 온실가스가 가장 적게 배출 되는 교통수단은?
  철도(5.9kg/인)
  ※ 항공 53.3, 버스 10, 승용차(휘발유) 50, 승용차(하이브리드) 38

- 생애주기 배출량이 가장 적은 생수 (500mL, PET)는?
  크리스탈 생수 (0.074kg/개)
  ※ e블루 0.08, 평창수 0.1, 삼다수 0.12
  ※ 출처 : 환경성적표지 유효 인증현황('21.5월 기준, 한국환경산업기술원)

온실가스 1톤의 의미를 알면, 우리 생활에서의 온실가스를 줄일 수 있습니다.

강 의 노 트

1.

2.

3.

제3장
탄소중립 에너지 기술 1

MS 해저 Data Center

◆ 서버 열
   : 심해 자연 냉각

◆ 데이터 입출력 전력
   : 조력 및 파력 이용

→ 안정적인 공기흐름으로 고장율이 지상 대비 1/8 수준

Source : 조선일보 (20210709)

우리나라 탄소중립 목표

- 2030년까지 2018년(727.6백만톤) 대비 40% 감축(291백만톤), 2050년까지 탄소중립
  - NDC 및 LEDS를 통해 국제사회에 우리나라 온실가스 감축목표 및 탄소중립 목표 선언
  - ('21.10.18) 2050년까지 국내 순배출량을 '0'으로 하는 2개 시나리오 발표

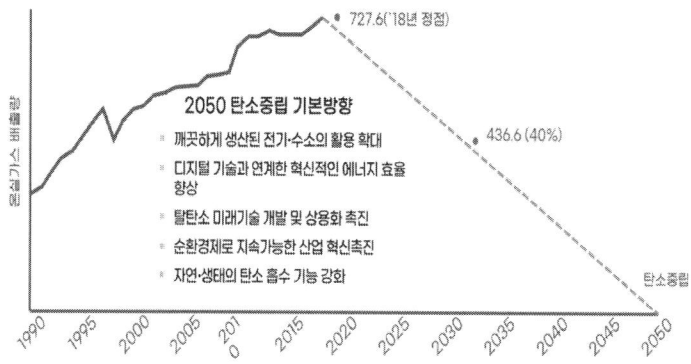

NDC : Nationally Determined Contribution(국가 자발적 기여), LEDS : Long-term greenhouse gas Emission Development (장기 저탄소 발전전략)

우리나라 탄소중립 Scenario

2050 탄소중립 시나리오 세부내용
(2021.10.18, 관계부처 합동)

- 2050 탄소중립 시나리오 A안
  - 화력발전 전면 중단 등 배출 자체를 최대한 줄이는 안
    - 2050년 총 배출량: 80.4백만톤
    - 총 흡수 및 제거량: 80.4백만톤(흡수원: 25.3백만톤, CCUS:55.1백만톤)
- 2050 탄소중립 시나리오 B안
  - 화력발전이 잔존하는 대신 CCUS 등 제거기술을 적용활용하는 안
    - 2050년 총 배출량: 117.3백만톤
    - 총 흡수 및 제거량: 117.3백만톤(흡수원: 25.3백만톤, CCUS: 84.6백만톤, 직접공기포집: 7.4백만톤)
- 주요 특징
  - IPCC 1.5℃ 특별보고서(`18)를 토대로 모든 국가가 2050년 탄소중립을 추진한다는 전제 下 국외 감축분이 없는 2050년 가정
  - 에너지전환에 대한 명확한 신호를 위해, 화력발전을 전반적으로 중단 혹은 최소화하는 방향으로 2050 탄소중립 시나리오 설정

탄소중립 10대 핵심기술 개발전략 (1/2)

태양광/풍력
- 태양전지 효율
  (現) 27%(상용 20) ⇒ (30) 35% ⇒ (50) 40%
- 풍력 발전기 용량
  (現) 5.5MW ⇒ (30) 15MW ⇒ (40) 20MW

수소
- 수소충전소 공급가(원/kg)
  (現) 7,000 ⇒ (30) 4,000 ⇒ (40) 3,000
- 수소 발전단가(원/kWh)
  (現) 250 ⇒ (30) 141 ⇒ (40) 131

바이오에너지
- 바이오연료 가격경쟁력 (동종 화석연료 대비)
  (現) 120~150% ⇒ (30) 100% ⇒ (45) 85%

수송효율
- 차세대전지 배터리 밀도
  (現) 250Wh/kg ⇒ (45) 600Wh/kg(상용화)
- 수소 고속충전기술
  (現) 1.6kg/분 ⇒ (30) 7.2kg/분

건물효율
- 건물 에너지 효율
  (30) 30% 향상 기술 확보
- 제로에너지 건축비(리모델링 대비)
  (現) 130% ⇒ (45) 105%

디지털화
- 데이터센터 전력 소모
  (30) 20%이상 저감
- 계통 운영시스템 적용
  (40) AI기반 차세대 계통 운영시스템 적용

CCUS
- CO₂ 상용급 포집 가격경쟁력
  (現) 60$/톤 ⇒ (30) 30$/톤 ⇒ (50) 20$/톤
- CO₂ 전환제품 가격경쟁력(기존 시장가 대비)
  (現) 연구중 ⇒ (40) 100%

Source : 탄소중립 기술혁신 추진전략, 에너지기술원

탄소중립 10대 핵심기술 개발전략 (2/2)

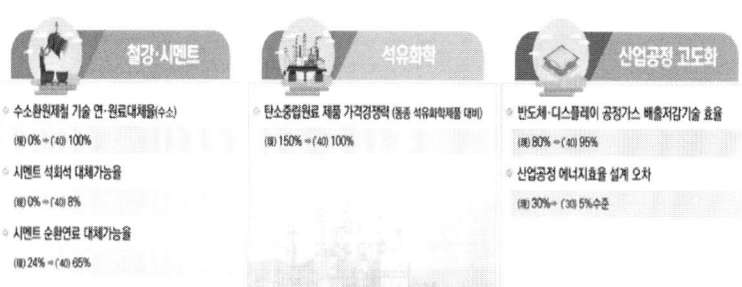

에너지 전환 부문 : 석탄발전 중단, 재생에너지 확대, 수소기반 발전
산업 부문 : 철강 / 석유화학-정유 / 시멘트 산업의 에너지 대체 및 효율화
수소 부문 : 수소 생산 탄소 Zero 기술 개발, 수소 생산 단가 저감
CCUS 부문 : 이산화탄소 포집-저장-활용 기술 개발

Source : 탄소중립 기술혁신 추진전략, 에너지기술원

2050 탄소중립 에너지 기술 Roadmap (1/3)

Source : 산업통상자원부 20211202 발표자료

2050 탄소중립 에너지 기술

➢ 주요 13대 분야 및 탄소 중립 산업-에너지 R&D 전략

13대 분야	청정 연료 발전	태양광	전력 계통	에너지 저장	그린 수소	산단 건물	자원 순환	CCUS	에너지 설비	정유
	연료 전지	풍력	섹터 커플링							
해당 분야	무탄소 발전	재생 에너지	계통 선진화	에너지 저장	수소화	에너지 효율화	자원 순환	CCUS	산업 공통 설비	정유

에너지 분야   공통 분야   산업 분야

Source : 산업통상자원부 20211202 발표자료

에너지 기술 관련 용어 중에서

➢ 섹터 커플링 (Sector Coupling)
: 인프라와 저장 가능한 에너지 (전력, 열, 수소)를 통해 발전, 난방 및 수송 부문을 연결하는 시스템

→ 발전부문의 잉여전력을 다른 부문에 활용

❖ P2X 로 표현
- P2G : Power to Gas
- P2L : Power to Liquid
- P2H : Power to Heat
- P2C : Power to Cooling
- P2M : Power to Mobility

Sector Coupling개념도

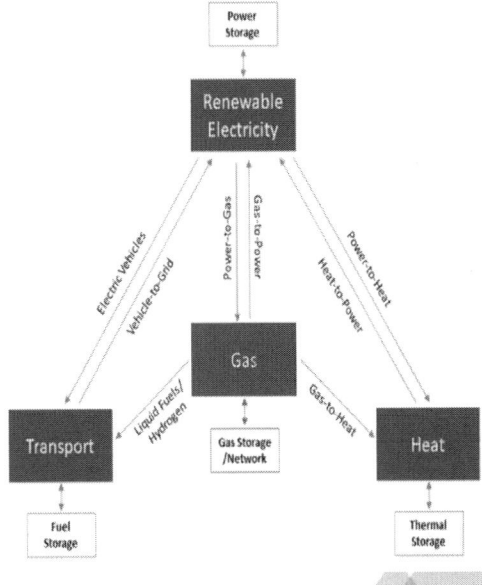

Source : Sterner et al (2014)

Sector Coupling을 이용한 에너지 효율

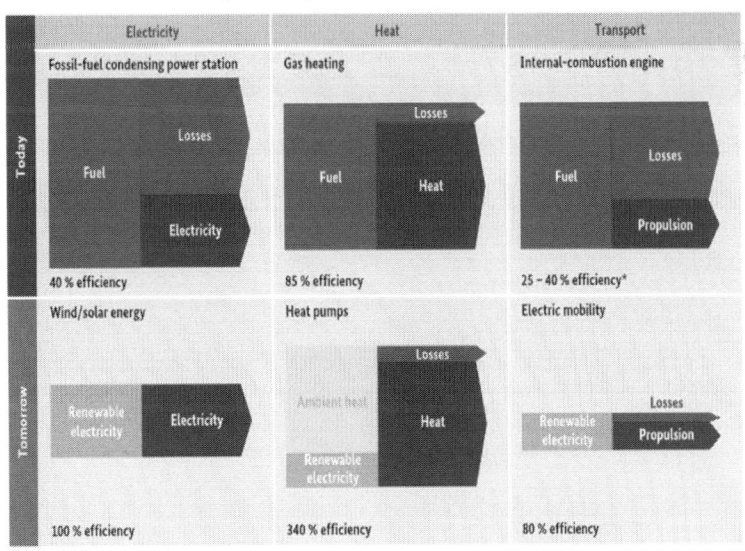

Source : An Electricity Market for Germany's Energy Transition, BMWi(2015)

제주도 CFI (Carbon Free Island) 2030

2030 탄소 없는 섬 제주도,
출력제한 없는 섬에서부터:
재생에너지 출력제한 문제 해결방안의 비용 분석

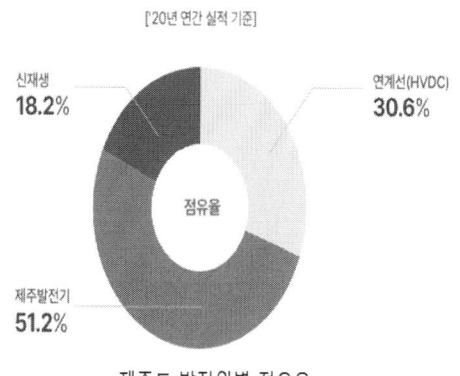

[20년 연간 실적 기준]

- 신재생 18.2%
- 연계선(HVDC) 30.6%
- 제주발전기 51.2%

제주도 발전원별 점유율

● 제주발전기(중앙급전 45.6%+기타(폐기물)5.6%)
● 신재생 세부 점유율(풍력 10.5%, 태양광 7.5%, 기타 신재생 0.2%)

제주도 CFI (Carbon Free Island) 2030

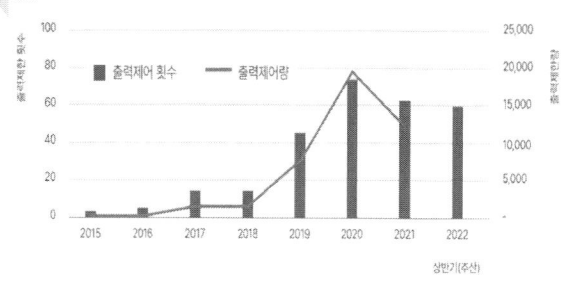

제주도 풍력발전 출력제한 현황   제주도 풍력발전 출력제한 사례

◆ 출력제한 : 수급 균형과 계통안정도를 유지하기 위한 조치
◆ 출력제한 안되면, 전력의 공급과 수요가 일치되지 않아 수급불균형이 발생하거나, 출력량이 급증하면서 주파수와 전압이 불안정하여 계통안정도에 영향을 주어 대규모 정전사태가 발생할 수 있음

Source : 한국전력거래소, 월별 제주 신재생에너지 발전 제어량

제주도 CFI (Carbon Free Island) 2030

➤ 출력제한 주요 원인

- ◆ 전력수요 고려 시, 상시가동 발전기와 연계선의 설비용량을 우선한 전력계통으로 유연성 낮음
- ◆ 재생에너지 초과공급량을 활용하는 에너지저장장치(ESS) 도입 미비
- ◆ 초과 생산된 전력의 역송전에 대한 기술 검토 미비

제주도 ESS 전력시장 개설
(8/19일 산업통상자원부 입찰공고)

➤ Implication

- ◆ Sector Coupling을 이용한 재생에너지 활용 극대화 가능성
- ◆ 정책 수립 시, 탄소중립에 대한 포괄적이고 장기적인 관점에서 접근 필요

국내 태양광 확대로 전력계통 안정성 저하

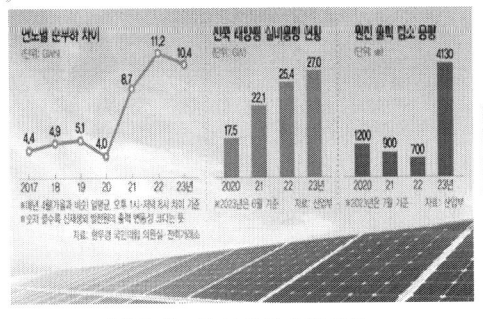

- ◆ 일몰 후 태양광 발전량 급하
  - → 다른 발전설비의 출력을 빠르게 올려 대응
  - → 출력변동성 증대 및 전력계통 안정성 저하

태양광 연도별 부하 및 용량 변화

대응방안

- ◆ 태양광 발전에 따른 전력수급 계량화 확대 : 자가발전량 계측 및 전력수급 예측도 향상
- ◆ 전력수급 기본계획에 탄소중립과 전력계통 안정성을 동시에 고려한 에너지 믹스 고려

Source : 서울경제신문 (20230820)

Renewable Energy Source

- 태양광, 태양열
- 풍력
- 바이오 에너지
- 소수력 에너지
- 해양 에너지
- 지열

Renewable Energy 장단점

ADVANTAGES	DISADVANTAGES
Renewable energy won't run out	Renewable energy has high upfront costs
Renewable energy has lower maintenance requirements	Renewable energy is intermittent
Renewables save money	Renewables have limited storage capabilities
Renewable energy has numerous environmental benefits	Renewable energy sources have geographic limitations
Renewables lower reliance on foreign energy sources	Renewables aren't always 100% carbon-free
Renewable energy leads to cleaner water and air	
Renewable energy creates jobs	
Renewable energy can cut down on waste	

Source : news.energytsage.com (2022. Nov.)

참고 : Energy Transition Index to RE

RE : Renewable Energy

Source : Fostering Effective Energy Transition (2023)

우주로 눈 돌리는 기후변화 대응

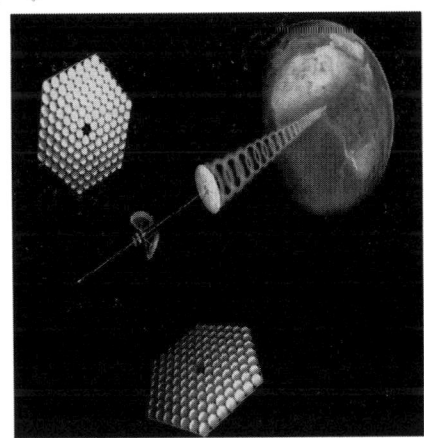

우주 태양광 발전 (SPS Type III 개념도)
(SPS : Space Solar Power)

Source : Spaceworks Engineering Inc.

❖ 우주 태양광 발전의 최고 장단점
- 장점 : 지표최고조건 지역대비 7배 효율
- 단점 : 매우 가혹한 우주환경으로 인해 발전 패널의 성능이 지표면 대비 8배 속도로 저하

❖ 2023.6.18 동아 사이언스 기사

미국 캘리포니아공대 연구팀이 우주에서 얻은 태양광 에너지를 마이크로파로 변환해 전기신호를 지구에 수신하는 데 성공. 우주에서 만든 태양광 에너지가 지구로 무선 전송된 첫 사례

우주로 눈 돌리는 기후변화 대응

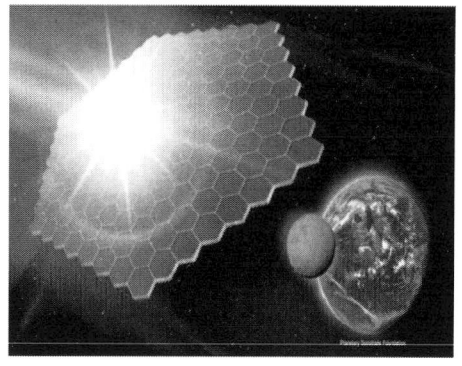

우주 차양막 개념도

Source : Planetary Sunshade Foundation

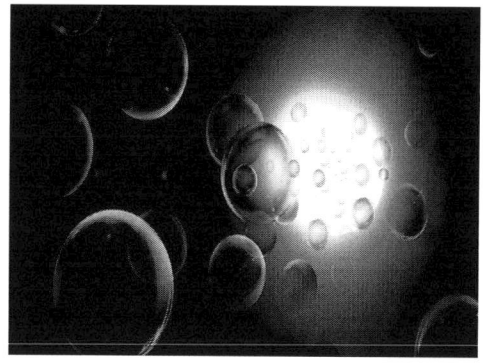

Space Bubble Raft 개념도

Source : space-bubbles@mit.edu

탄소 중립 연료 (e-Fuel)

탄소중립 연료 (e-Fuel)

> 탄소 중립 연료란?

◆ 전기 기반 연료 (Electricity-based fuel)로 e-Fuel (or Electrofuel)이라 표현

1) 재생에너지 전기로부터 얻어진 그린수소*로부터 합성된 탄소중립 연료를 의미
   → **그린수소** : 100% 재생에너지를 이용해 전기를 생산하고, 이 전기를 기반으로
                수소를 생산하는 과정에서 탄소배출이 없는 수소

2) 재생에너지를 수전해를 통해 그린수소 및 파생된 e-Fuel (그린 암모니아 포함)
   → **그린암모니아** : 100% 재생에너지를 활용하여, 온실가스 배출이 없이
                   생산되는 암모니아

탄소 Cycle 비교

> 화석연료 기반 (a)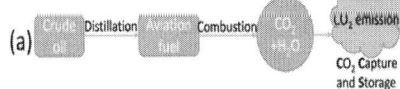

> 탄소중립 연료 (b)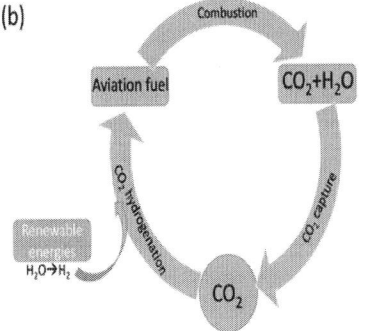

<u>탄소 Cycle 모식도</u>
(a) open carbon cycle, b) closed carbon cycle)

Source : Nature Comm., 11(1) (2020)

e-Fuel 합성 경로

Chemical	Pathway	
Green hydrogen	Water electrolysis with renewable energy	
Green ammonia	(Feedstocks: $H_2 + N_2$) 1. Haber-Bosch process 2. Electrochemical process	
Methanol	(Feedstocks: $H_2 + CO_2$) 1. CO_2 hydrogenation 2. RWGS → CO hydrogenation	RWGS : Reverse Water Gas Shift
Methane	(Feedstocks: $H_2 + CO_2$) Sabatier reaction	
DME	(Feedstocks: $H_2 + CO_2$) Methanol dehydration	DME : Di-Methyl Ether
Syncrude	(Feedstocks: $H_2 + CO_2$) 1. RWGS → CO-FT 2. SO co-electrolysis → CO-FT 3. CO_2-FT 4. Methanol-to-Gasoline	FT : Fisher-Tropsch Synthesis

Source : Jr. Korean Soc., Combust., 27(1) (2022)

연료들의 물성 비교

> Properties of various fuels @ 300K, 1atm

	Gasoline	Methane	Hydrogen	Ammonia
Composition	$C_4 - C_{12}$	CH_4	H_2	NH_3
Density [kg/m³]	700	0.651	0.082	0.73
Specific energy [MJ/kg]	42.5	50	120	18.8
Minimum autoignition temperature [°C]	300	630	520	650
Adiabatic flame temperature [°C]	1950	1950	2110	2000
Maximum laminar burning velocity [m/s]	-	0.37	2.91	0.07
Flammability limit (volume of air) [%]	1.4-7.6	4.4-16.4	4-75	15-28

Source : Proc, Combust. Inst. 37(1) (2019)

대체 연료들의 장단점

	Green hydrogen	Green ammonia	E-Fuel	Biofuel
Feedstock	Water	Hydrogen, nitrogen	Hydrogen, carbon dioxide	Freshwater, energy crop/algae
Applications	- Carbon free combustion - Fuel cell feedstock	- Carbon free combustion - Fuel cell feedstock - Hydrogen carrier	- Carbon neutral combustion	- Carbon neutral combustion
Advantages	- Additional synthesis processes not required - High gravimetric energy density - Broad flammability range	- Existing transport and storage infrastructure - Mature production technology	- Drop-in fuel - Wider range of carbon number compared to biofuel	- Drop-in fuel - Dispatchable generation - Biodiesel-blended fuel is currently commercialized
Disadvantages	- Low volumetric energy density - Difficult to store and transport - Engine modification required for a sole use	- Engine modification required for a sole use - Toxic to human - Corrosive to copper, plastics, etc. - NOx exhaust - Metal embrittlement	- Low energy efficiency compared to direct use of electric energy - Currently not viable economically	- Drinking water required as a feedstock - Environmental impact of biomass production is not studied sufficiently.

Source : Int. Jr. Hydrogen Energy 43(45) (2018)

e-Fuel 장단점

❖ 장점

- 기존 내연기관에 적용 가능
- 기존 화석연료 network 활용 가능
- 포집된 이산화탄소를 원료로 사용하므로 탄소중립에 기여

❖ 단점

- 신규 투자 필요, 경제성 낮은 수준
- 규모경제 도달 수준 미흡

Liquid e-Fuel

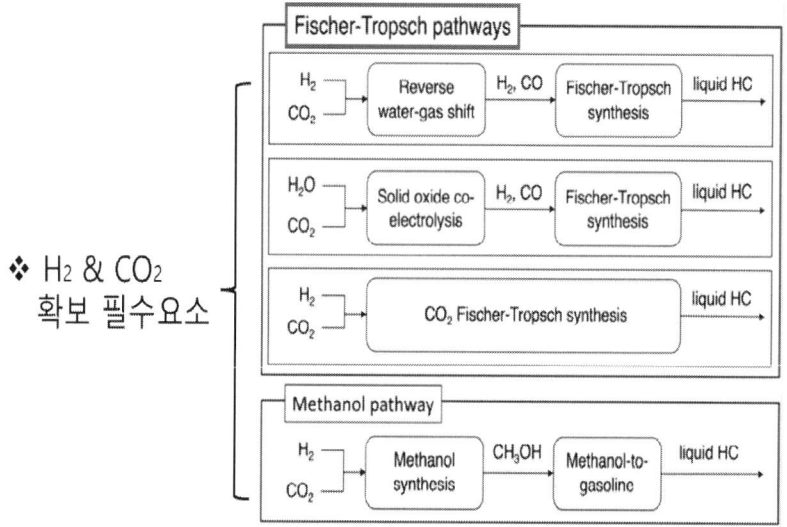

❖ H₂ & CO₂ 확보 필수요소

RWGS & FT Reaction Pathway

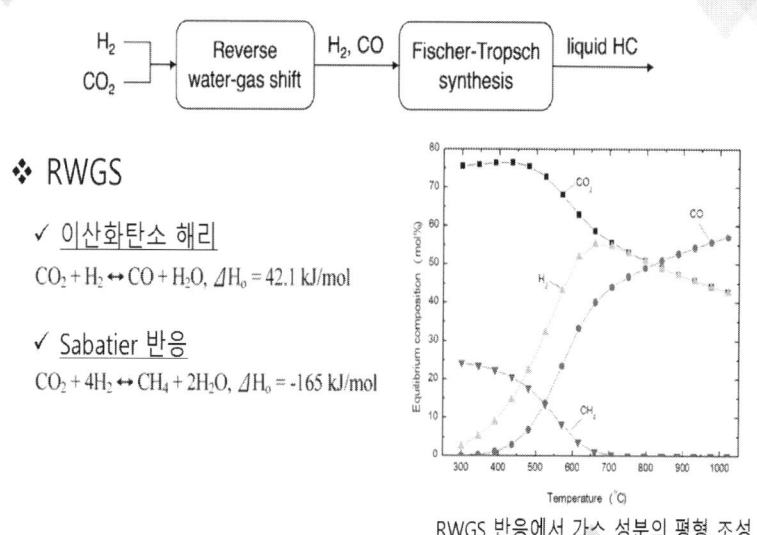

❖ **RWGS**

✓ <u>이산화탄소 해리</u>
$CO_2 + H_2 \leftrightarrow CO + H_2O,\ \Delta H_o = 42.1\ kJ/mol$

✓ <u>Sabatier 반응</u>
$CO_2 + 4H_2 \leftrightarrow CH_4 + 2H_2O,\ \Delta H_o = -165\ kJ/mol$

<u>RWGS 반응에서 가스 성분의 평형 조성</u>

Source : Jr. Climate Change Res., 16(1)(2021)

RWGS & FT Reaction Pathway

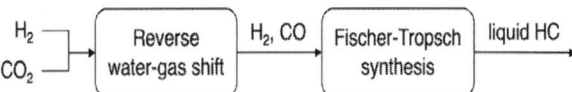

❖ FT Synthesis

✓ Chain Growth Reaction
 $nCO + 2nH_2 \rightarrow (-CH_2-)_n + nH_2O$

✓ Product distribution eq*.
 $x_n = (1-\alpha)\alpha^{n-1}$

* ASF distribution : Anderson-Schulz-Flory distribution
 x_n : Specific Carbon Number (mole ratio)
 α : Chain growth probability
 n : Carbon Number

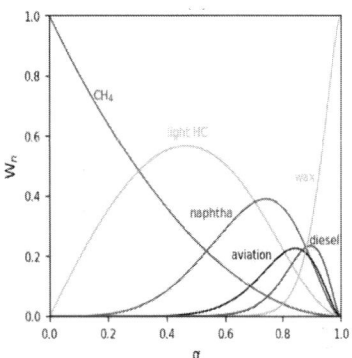

FT 반응의 반응물 분포

Source : FT Refining (Wiley-VCH Verlag & Co. Germany) (2011)

RWGS & FT Reaction Pathway

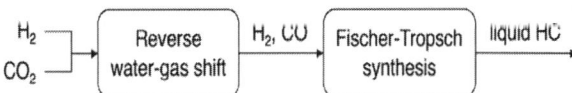

❖ FT Synthesis

✓ Chain Growth Reaction
 $nCO + 2nH_2 \rightarrow (-CH_2-)_n + nH_2O$

✓ Product distribution eq*.
 $x_n = (1-\alpha)\alpha^{n-1}$

* ASF distribution : Anderson-Schulz-Flory distribution
 x_n : Specific Carbon Number (mole ratio)
 α : Chain growth probability
 n : Carbon Number

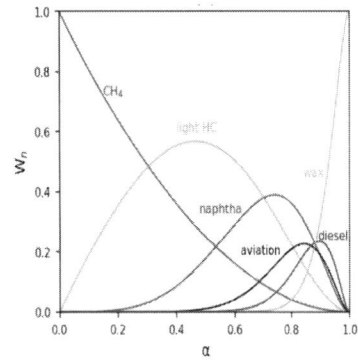

FT 반응의 반응물 분포

Source : FT Refining (Wiley-VCH Verlag & Co. Germany) (2011)

CO_2를 이용한 FT Reaction Catalysts

Active Metal	Promoter		Support	
	Precious Metal & Rare Earth Metal	Transition Metal & Alkali Metal	Oxide	Zeolite
Fe, Co, Ni, Os	Pt, Pd, Ru, Re, La, Ce	Na, K, Ca, Cu, Mn, V, Ta, Zr, Mo, Zn, Cr	Al_2O_3, TiO_2, SiO_2, ZrO_2, MgO	ZSM-5, MCM-22

- FT Reaction 반응 온도 및 압력 ↓
- Conversion & Selectivity ↑
- Catalyst Stability ↑

Source : Energy Fuels 35 (2021)

e-Fuel Cost vs Fossil Fuel Prices

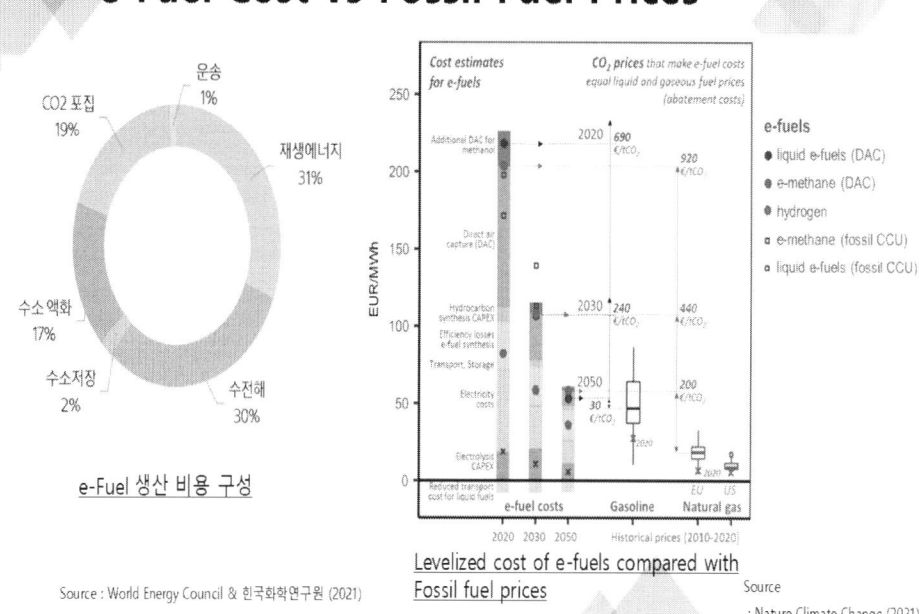

e-Fuel 생산 비용 구성

Source : World Energy Council & 한국화학연구원 (2021)

Levelized cost of e-fuels compared with Fossil fuel prices

Source : Nature Climate Change (2021)

CO_2에 따른 e-Fuel 가격 예상

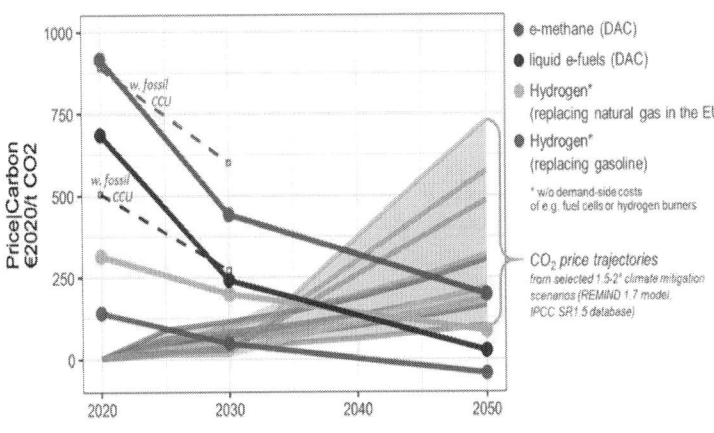

Trajectories for required CO_2 prices to make e-fuels competitive with fossil fuels

Source : Nature Climate Change (2021)

e-Fuel 공급 가격 전망 (EU)

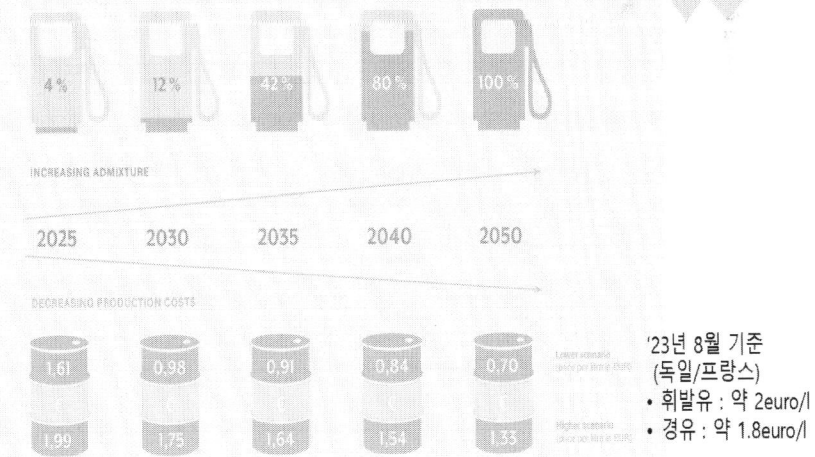

Source : Prognos AG, Germany (2018)

'23년 8월 기준 (독일/프랑스)
- 휘발유 : 약 2euro/l
- 경유 : 약 1.8euro/l

제4장

탄소중립 에너지 기술 2

Green H₂ & H₂ Economy

우주 vs 지구의 대기 구성 성분

우주 대기 구성 성분

순위	물질	비율(%)
1	수소	73.900
2	헬륨	24.000
3	산소	1.040
4	탄소	0.460
5	네온	0.130
6	철	0.100
7	질소	0.090
8	규소	0.065
9	마그네슘	0.058
10	황	0.044

- 생물을 구성하는 주요 성분
- 질소 78 %
- 산소 21 % — 호흡에 사용
- 아르곤 0.93 %
- 기타 0.04 %
- 이산화 탄소 0.03 % (300ppm) — 광합성, 온실 효과를 일으킴

❖ 지구 대기 중 수소 농도
 : 500~550ppb (0.5~0.55ppm)

지구 대기 구성 성분

Source : www.ibric.org/myboard (2021)

Green H₂의 활용

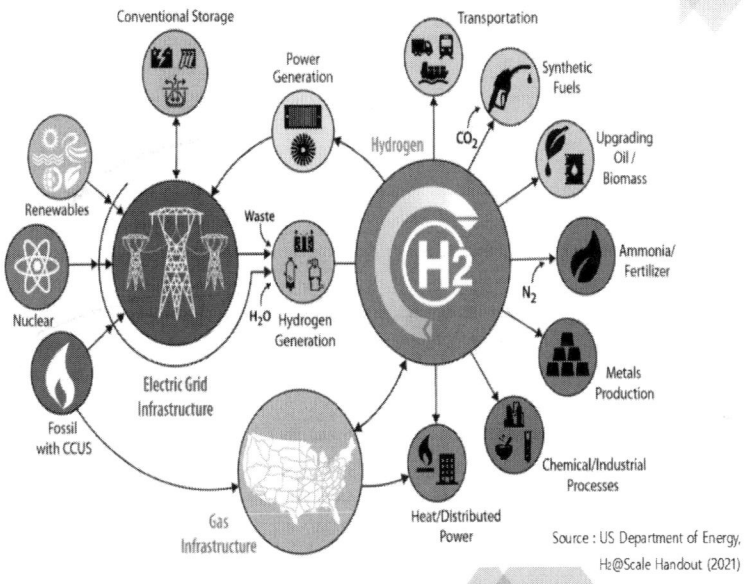

Source : US Department of Energy, H₂@Scale Handout (2021)

수소의 분류

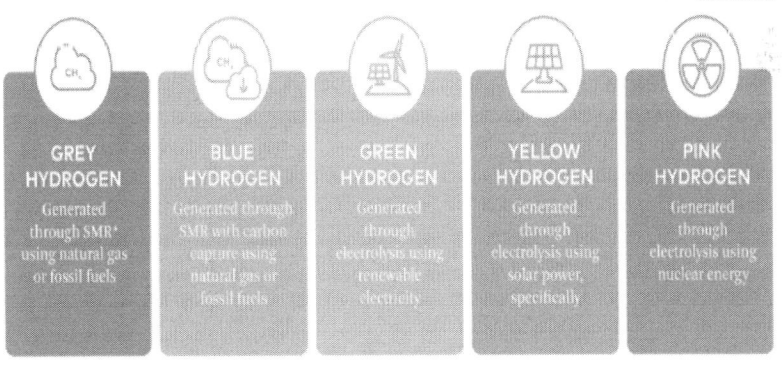

* 최근에 Turquoise Hydrogen (청록 수소) : Methane Pyrolysis Technology

수소 생산 방법 및 Source

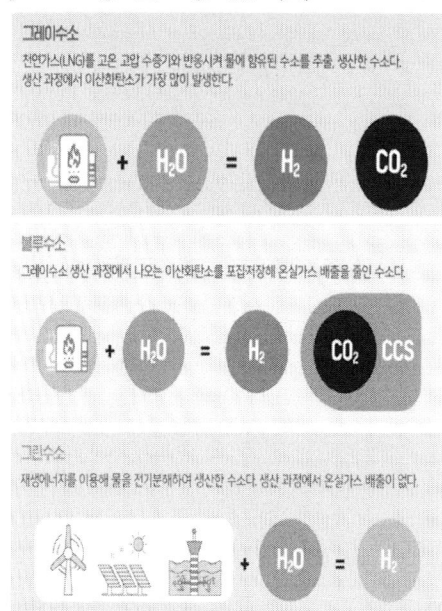

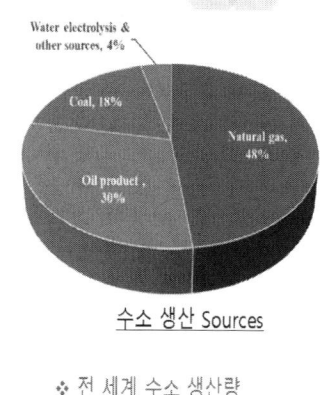

수소 생산 방법

수소 생산 기술 분류

원료	에너지원 및 화학반응	
연료 이용 수소 생산	개질 반응	가스연료(LNG, LPG) 개질
		부생가스 개질
		합성연료(메탄올, DME) 개질
	가스화 반응	석탄/펫코크 가스화
	생물학적 전환 반응	생물학적 CO 전환
바이오매스/폐자원 이용 수소 생산	가연성 폐자원 가스화	가연성 폐자원 가스화
	바이오매스 가스화	바이오매스 가스화
	생물학적 발효	생물학적 발효
물분해 수소 생산	수전해(전기 분해)	알칼리 수전해(AEC)
		고분자전해질 수전해(PEMEC)
		고체산화물 수전해(SOEC)
	광 분해	광전기화학(PEC)
		광촉매
		광생물학
	열 분해	열화학사이클
		레독스사이클
	원자력	초고온가스로

Source : KISTEP 2021 Report

수소 생산 방법에 따른 장단점

구분	정의	에너지원	특징	국내 생산가격 ('18년)
부생수소	석유화학 공정이나 제철공정에서 화학반응에 의해 부수적으로 생산	주로 화석 연료	• 폐가스 활용 • 정제 필요 • 생산량 확대 한계	~2000원/kg
개질수소 (그레이수소)	화석연료를 활용하여 촉매 반응으로 생성	화석연료 (천연가스, 석탄)	• 대량생산 가능 • 저렴한 생산단가 • CO_2 발생 多	2,700 ~ 5,100원/kg
개질수소 + CCUS (블루수소)	개질수소 + CCUS 장치를 통해 발생된 CO_2 포집	화석연료 (천연가스, 석탄)	• CO_2 발생 小 • 포집된 CO_2의 효율적 활용방안 필요	-
수전해 (그린수소)	물분해방식으로 물에 전기를 가하여 수소와 산소 생성	신재생에너지	• CO_2 발생 無 • 높은 생산 단가 • 지역적 제한	9,000원 ~ 10,000원/kg

Source : KISTEP 2021 Report
(수소연료전지와 연관산업 기술개발 동향과 시장전망, 재구성 자료)

2040 국내 수소 가격 전망

➤ 수소 공급 및 가격 전망

Source : 산업통상자원부 2019년 발표 자료

	2018	2022	2030	2040
Supply	130K ton/yr	470K ton/yr	1,940K ton/yr	5,260K ton/yr
Method	① By-product(1%) ② Reforming(99%)	① By-product ② Reforming ③ Electrolysis	① By-product ② Reforming ③ Electrolysis ④ Import ※①+③+④: 50% ② : 50%	① By-product ② Reforming ③ Electrolysis ④ Import ※①+③+④: 70% ② : 30%
Price	-	₩ 6,000 /kg (Initial Market Price)	₩ 4,000 /kg	₩ 3,000 /kg

➤ 현재 전력용 연료 단가

Source : 한국전력, 전력통계정보시스템

기간	연료단가				
	원자력	유연탄	무연탄	유류	LNG
	원/KWh	원/ton	원/ton	원/kl	원/ton
2023 / 08	6	218,090	176,378	1,246,137	1,100,409

연료별 부피 및 무게당 에너지 밀도

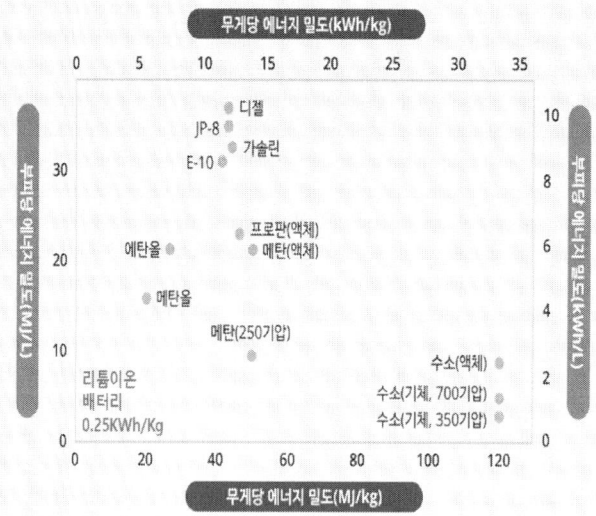

Source : US DOE 자료 (2019)

참고) 2차 전지 에너지 밀도

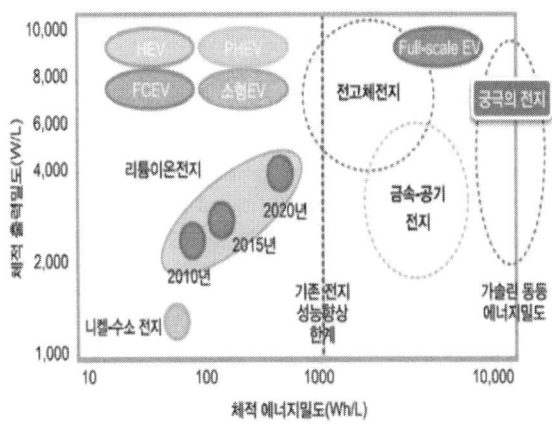

2차 전지 에너지 밀도 비교

Source : www.amenews.kr/m

수소 생산관련 기술 성숙도

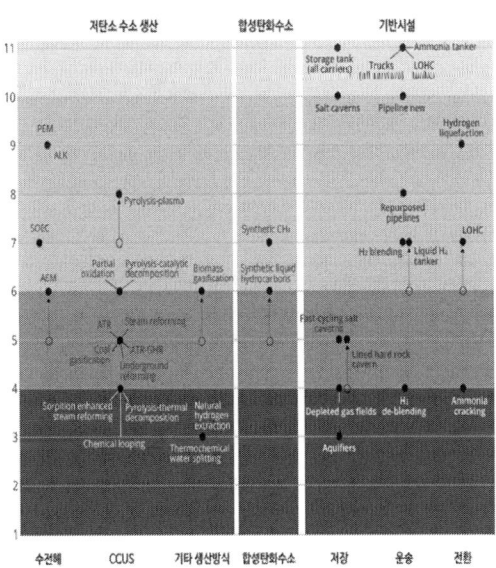

Source : IEA, Global Hydrogen Review (2022)

수소 활용관련 기술 성숙도

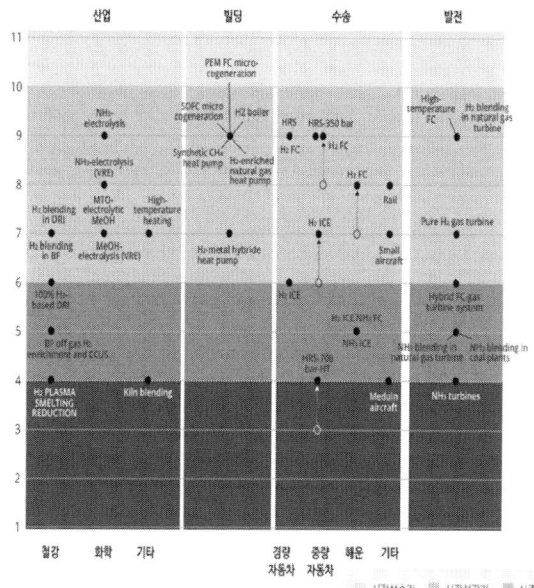

Source : IEA, Global Hydrogen Review (2022)

수소 활용부문 Roadmap (EU)

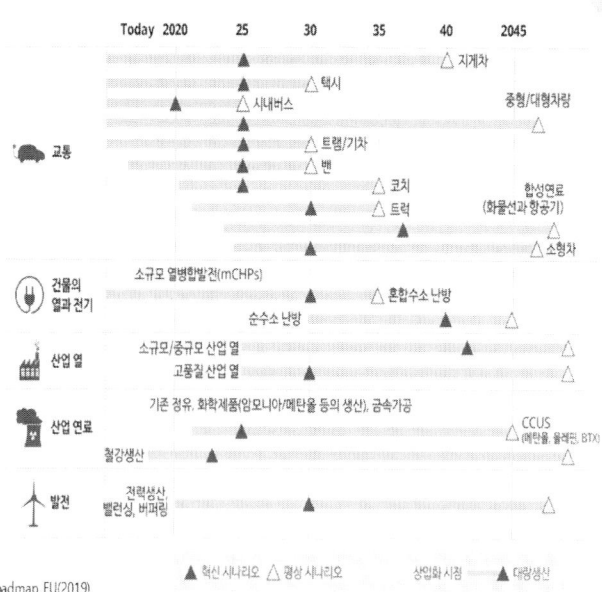

Source : Hydrogen Roadmap EU(2019)

탄소경제에서 수소경제로

구분	탄소경제	수소경제
에너지 패러다임	탄소자원(석유, 석탄, 가스 등) 중심	탈탄소화 수소 중심
	수입 의존(99%)	국내 생산으로 에너지 자립 기여
에너지 공급	대규모 투자가 필요한 중앙집중형 에너지 수급	소규모 투자로 가능한 분산형 에너지 수급
	입지적 제약이 크고 주민 수용성이 낮음	입지적 제약이 적고 주민 수용성이 높음
경쟁 양상	자원개발 및 에너지 확보 경쟁	기술경쟁력 확보 및 규모의 경제 경쟁
환경성	온실가스, 대기오염물질 배출	온실가스 배출이 적어 친환경적
	* CO_2, NO_x, SO_x	* 부산물 = (H_2O)

Source : 수소경제 활성화 로드맵, 관계부처 합동(2019)

수소경제 당위성과 해결해야 할 과제

❖ 당위성

- 지구온난화 대응을 위한 세계적 협약 및 환경규제 강화
- 정부의 강력한 정책적 의지
- 신규 활용 분야 등장 및 기술 개발 관심도 증가

❖ 단점

- 수소 생산, 저장 및 수송, 이동 등 수소 생태계의 기술 수준 미흡
- 대규모 투자 필수로 경제성 낮음

Source : 과학기술정책연구원 (2020)

국내 수소산업 생태계 구조 및 경쟁력

	수소생산	저장·운송	충전	수소모빌리티	수소연료전지
제품 및 방식	· 부생수소 · 추출수소 · 수전해수소	· 파이프라인 · 튜브트레일러 · 액화탱크로리	· 파이프라인 · 튜브트레일러 · 이동식 · 개질 · 수전해	· 수소차 · 수소버스 · 수소선박 · 수소열차 · 수소드론 등	· 수송용 · 가정·건물용 · 발전용 · 휴대용
부품 소재	· 수소 제조 장치 (개질, 전기분해) · 개질기, 탈황장치, 수전해장치, PSA, 압축기	· 수소저장용기 · 트레일러 · 수소공급배관 · 유량계, 촉매, 센서, 고압개관· 밸브 등	· 압축기, 고압용기, 디스펜서	· 연료전지시스템 (스택, 수소·공기 공급장치 등) · 수소저장장치 · 전장장치 · 운전장치	· 셀스택 · 연료변환기 · BOP · 전자장치
국내 성과	· 부생수소 상용화 · 추출·수전해는 실증단계	· 고압기체 상용화 · 액화·액상기술은 개발단계	· 수소충전소 31기 구축(2019.10)	· 수소차 양산단계 · 선박·열차·드론 등은 R&D 단계	· 수송용(수소차)· 발전용·건물용 수소 연료전지 분야 선도국 수준
경쟁력 수준	· 원천기술 미흡 · 상용화·실증 부족 (추출,수전해)	· 핵심기술 미흡 · LNG·부생수소 파이프라인 구축 경험	· 부품 국산화율 미흡(40%) · 사업성 부족	· 수소차 양산 (2013) · 부품 국산화 99% · 소재기술 미흡	· 시스템·운영기술 확보 · 전극, 촉매, 전해질 수입의존

Source : 산업연구원, 한국 수소산업 생태계 분석과 발전과제 (2019)

참고) 전기차 vs 수소차

전기차 vs 수소차 비교	전기차(EV)	평가	수소차(FCEV)
친환경성	오염물질 미배출	<	물만 배출, 미세먼지 정화 기능
주행거리	350~400km	<	600km 이상
충전시간	현대차 코나 일렉트릭 급속 20~30분	<	3~5분 현대차 넥쏘
유지비	1km당 25원	>	1km당 73원
충전소 설치비	1억원	>	30억원
최대출력	평균 200마력	>	평균 150마력
가격	4650만원(코나 엘렉트릭)	>	6890만원(넥쏘)

Source : 신문기사, 서울신문 (2019)

수소의 저장 방법

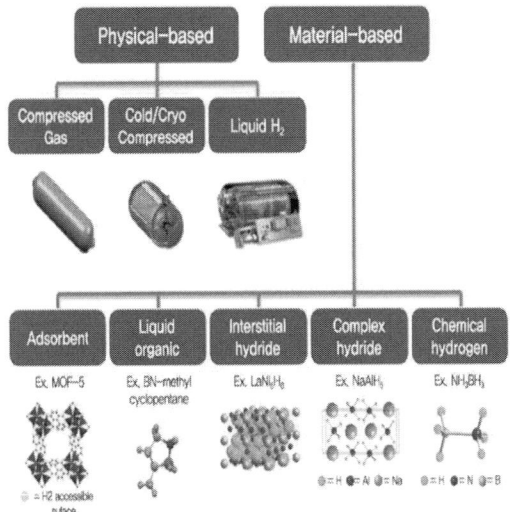

Source www.energy.gov/eere/fuelcells/hydrogen-storage

수소의 저장 방법 및 저장량

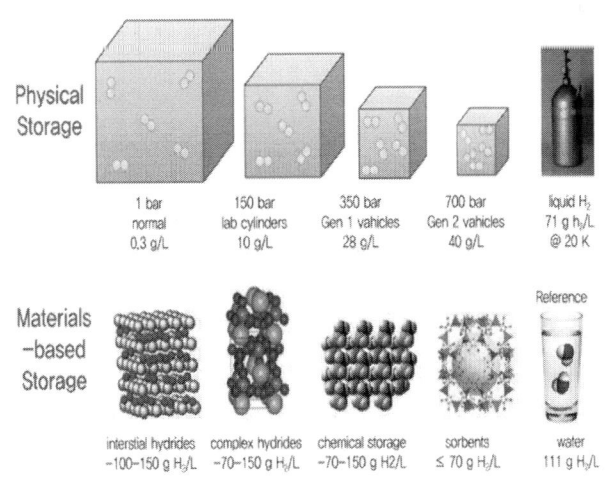

Source Int. Jr. Hydrogen Energy 42(1) (2017)

수소의 저장 방법의 장단점

고압기체 저장기술	액화 저장기술	고체 저장기술	화학적 저장기술
고압(~900Bar) 기체상태로 저장(상용화 기술)	-253℃ 극저온에서 액화 상태로 저장(소규모실증)	다공성 물질 표면에 흡착 저장(연구개발)	유기화합물 수소저장기술 (연구개발)
장점	**장점**	**장점**	**장점**
• 현재 가장 보편화 기술 • 수소차 및 버스에 적용 가능	• 상압(3Bar)에서 저장 가능 • 체적당 에너지저장 밀도 높음	• 체적당 에너지저장 밀도 높음 • 안전성이 높음	• 상온, 고압에서 취급 • 대용량 수소저장 가능
단점	**단점**	**단점**	**단점**
• 고압저장시 대량에너지 필요 • 체적당 에너지저장 밀도 낮음	• 수소기화손실 발생(장기보관X) • 액화시 대량에너지 필요	• 중량당 저장밀도 낮음(무거움) • 수소저장합금 고가(경제성X)	• 유기화합물 재생공정 필요 • 수소화, 탈수소화 설비 필요

> 수소 생산 및 이송 추정 비용

단위 : USD/Kg

	액체수소	파이프라인	튜브 트레일러
생산 비용	2.21	1.00	1.30
이송 비용	0.18	2.94	2.09
총 비용	3.66	5.00	4.39

Source KIST, 미래에셋 리서치센터 (2020)

액화수소 Project (일본)

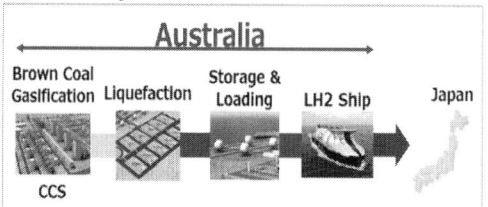

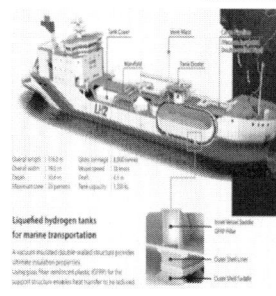
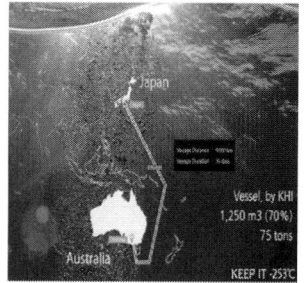

Source : Review of H2 Transport and Its Perspective (Liquefied H2), 2020, HySTRA Project

액화수소의 대안? → LOHC* & NH₃

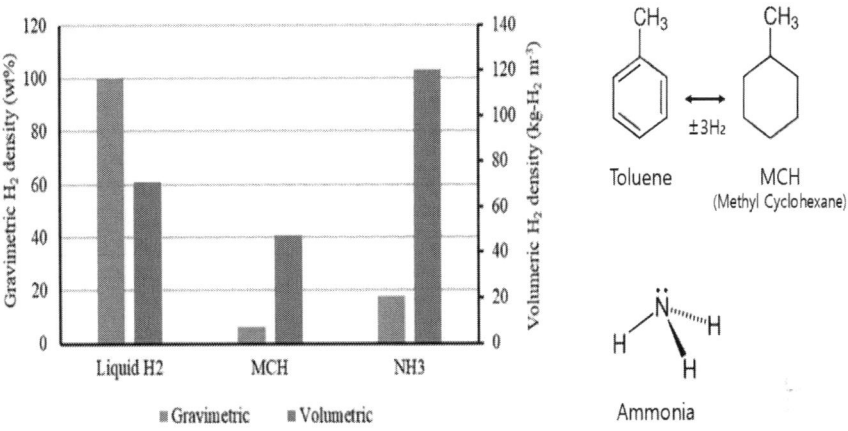

* LOHC : Liquid Organic Hydrogen Carrier

Source : Int. Jr. H₂ Energy 44 (2019)

대표적인 LOHC

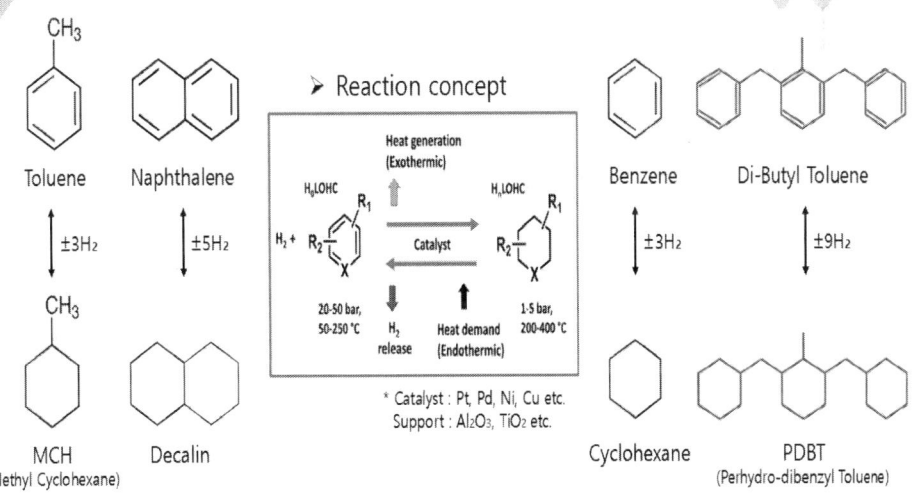

Source : Int. Jr. H₂ Energy 44 (2019) & Energies 13 (2020)

Ammonia Cracking 기술 수준

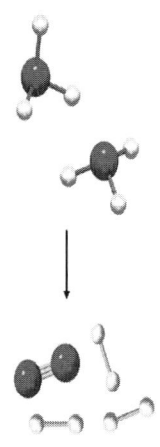
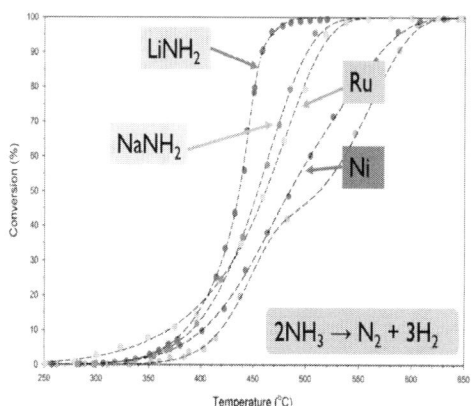

$2NH_3 \rightarrow N_2 + 3H_2$

Source : Univ. Birmingham 강의 자료 (2021)

액화수소와 Ammonia

항목	액화수소	암모니아
부피	기체(수소) 대비 부피 800분의1	기체 수소 대비 부피 1400분의1
장점	탈수소화·정제공장 불필요 에너지 손실 적음 기존 LNG 터미널 재활용 가능	높은 에너지 밀도, 암모니아 액화(-35℃) 비용 상대적 저렴, 기존 프로판 기반 시설 이용 가능
단점	대규모 시설 비용, 극저온(-253℃) 유지·저장, 증발가스 발생, 기술 장벽	정제 시 고열(600℃ 이상) 및 에너지 필요, 독성과 고약한 냄새
주요 기업	두산에너빌리티, SK E&S, 효성중공업, 한화오션	한국전력, 현대자동차, 포스코홀딩스, 롯데케미칼

Source : 서울신문

➤ Properties of various fuels @ 300K, 1atm

	Gasoline	Methane	Hydrogen	Ammonia
Composition	C_4 - C_{12}	CH_4	H_2	NH_3
Density [kg/m^3]	700	0.651	0.082	0.73
Specific energy [MJ/kg]	42.5	50	120	18.8
Minimum autoignition temperature [℃]	300	630	520	650
Adiabatic flame temperature [℃]	1950	1950	2110	2000
Maximum laminar burning velocity [m/s]	-	0.37	2.91	0.07
Flammability limit (volume of air) [%]	1.4-7.6	4.4-16.4	4-75	15-28

Am 8.9배 ↑
Hy 6.4배 ↑

Source : Proc. Combust. Inst. 37(1) (2019)

Ammonia의 분류

구분		그레이 암모니아	블루 암모니아 (청정 암모니아)	그린 암모니아
생산방식		그레이수소로 제조	블루수소로 제조	그린수소로 제조 (재생 전력을 활용한 수전해 방식으로 수소를 생산한 후 공기 중 질소를 합성)
		화석연료에서 수소를 생산한 후 공기 중 질소를 합성		
온실가스	발생	수소 생산 시 발생	수소 생산 시 발생	발생하지 않음
	처리	대기 중 방출	CCS를 활용하여 포집하므로 대기 중 방출되지 않음	

Source 전기저널 20220524

암모니아의 생산 방법 및 Source

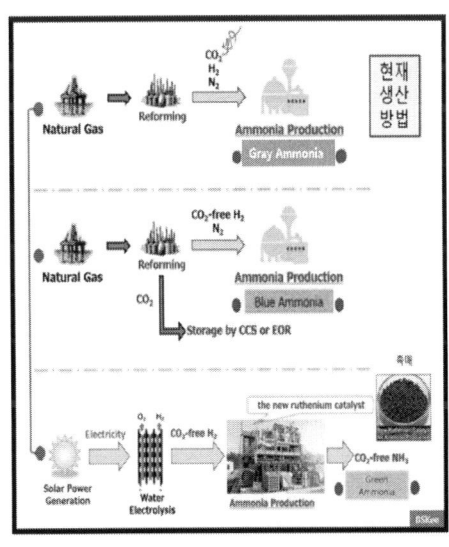

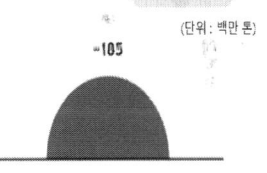

(단위 : 백만 톤)
~185
Fertiliser and existing industrial applications

'20년 암모니아 생산량 및 용도

❖ 전 세계 암모니아 생산량
- 1억8천5백만 톤 ('20 기준)
→ 100% Grey Ammonia

Source : IEA (2021)

Green Ammonia 생산 및 활용

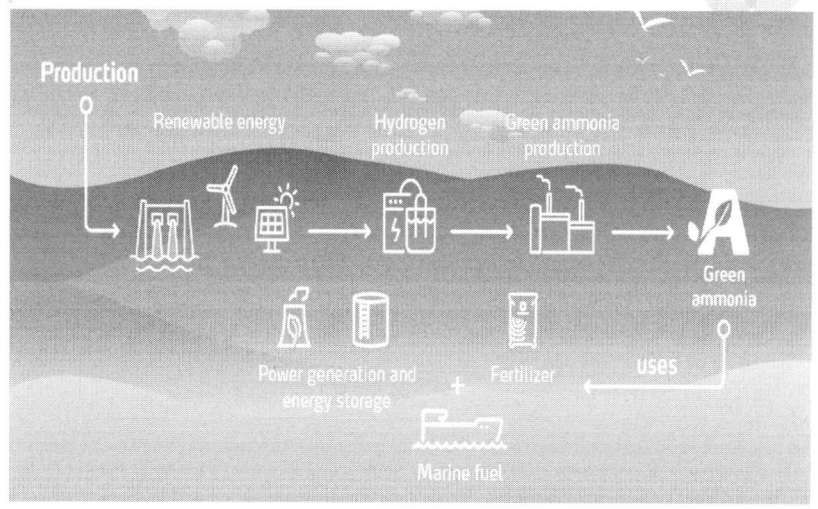

Source : World Economic Forum (2021)

Green Ammonia의 향후 용도

❖ 현재 대비 3~6배의 성장이 기대

(단위 : 백만 톤)

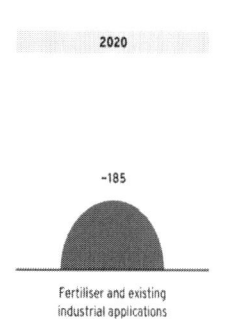
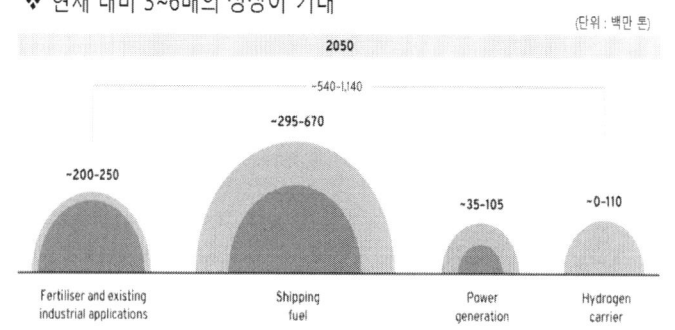

✓ 선박용 연료로의 활용
✓ Ammonia의 연료로 활용 시, De-NOx 설비 필수

Source : IEA (2021) & Ammonia Technology Roadmap : Food and Lans Use (2019))

참고) Ammonia 연료 선박용 엔진 개발

❖ 독일의 엔진 설계회사 (MAN Energy Solutions)
  연구용 암모니아 엔진의 연소 실험에 성공

❖ HD현대중공업 '24년까지
  첫번째 선박용 암모니아 연료
  엔진 납품 계획
  (HD현대중공업 :
   MAN의 엔진 생산 면허 보유)

연구용 암모니아 엔진 (MAN Energy Solutions)

Source : 조선비즈 (20230712)

Green 수소와 Green Ammonia의 발전 활용

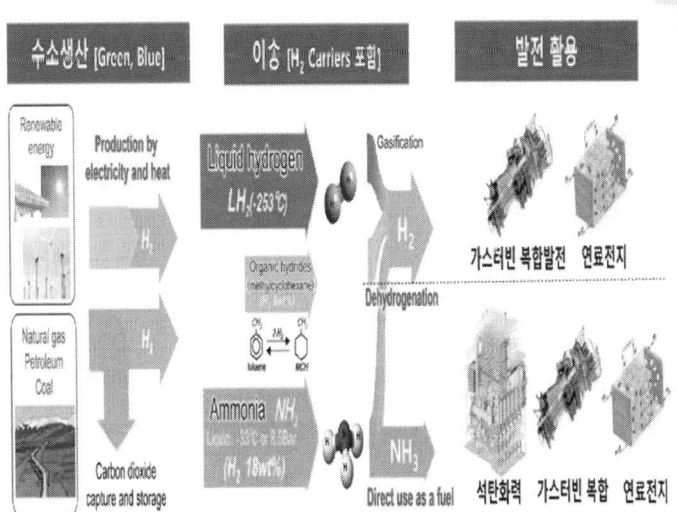

Source : 산업통상자원부 (2021)

참고) H₂, NH₃ & H₂+NH₃ Combustion

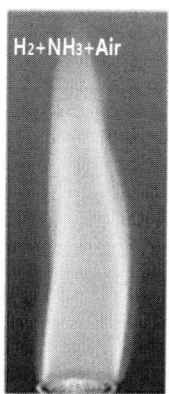

Source : KAIST, 항공우주공학과 실험 자료 (2022)
한국에너지기술연구원 발표 자료 (2021)
Energy Fuels 36(16) (2022)

액화수소, LOHC, Green NH3 예상 가격

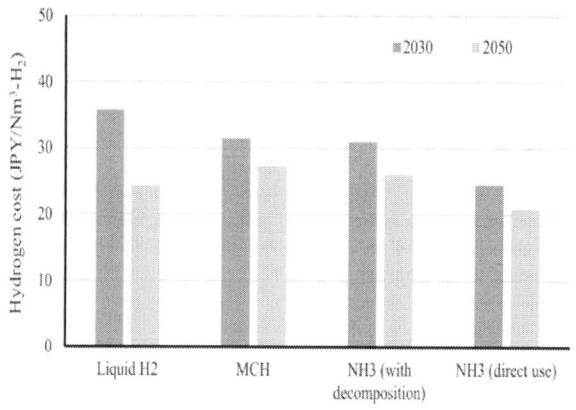

Source : Int. Jr. H2 Energy 44 (2019),
Economic Analysis on International Hydrogen energy carrier supply chains (2017)

Natural Hydrogen

We weren't looking in the right places with the right tools.

GEOFFREY ELLIS | U.S. GEOLOGICAL SURVEY

The hydrogen rainbow
Researchers use colors to distinguish between different kinds of hydrogen.

- **Gray hydrogen**
  Made from fossil fuels, which release carbon dioxide and add to global warming.
- **Blue hydrogen**
  Same as gray hydrogen, but the carbon is captured and sequestered.
- **Green hydrogen**
  Made without carbon emissions by using renewable electricity to split water.
- **Gold hydrogen**
  Tapped from natural subsurface accumulations.
- **Orange hydrogen**
  Stimulated by pumping water into deep source rocks.

Source : Science 16 Feb 2023

Natural Hydrogen

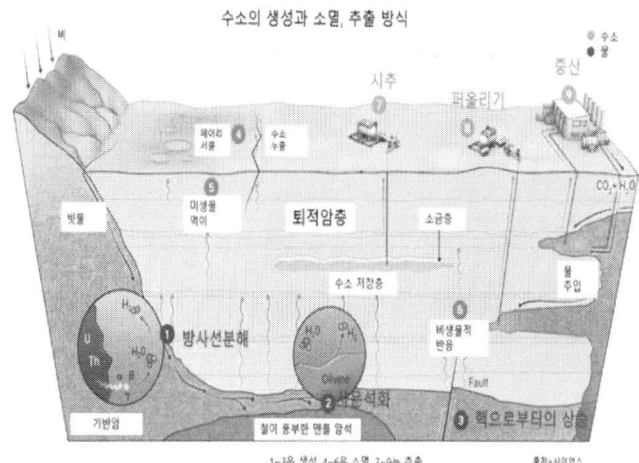

Source : Science 16 Feb 2023, 일부 편집

Natural Hydrogen

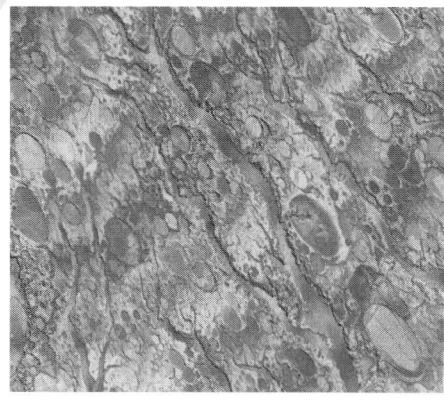

Source : Science 16 Feb 2023

Natural Hydrogen

He and his USGS colleague Sarah Gelman gave it a try using a simple "box" model borrowed from the oil industry. The model accounted for impermeable rock traps of different kinds, the destructive effect of microbes, and the assumption—based on oil industry experience—that only 10% of hydrogen accumulations might ever be tapped economically. Ellis says the model comes up with a range of numbers centered around a trillion tons of hydrogen. That would satisfy world demand for thousands of years even if the green-energy transition triggers a surge in hydrogen use.

Source : Science 16 Feb 2023

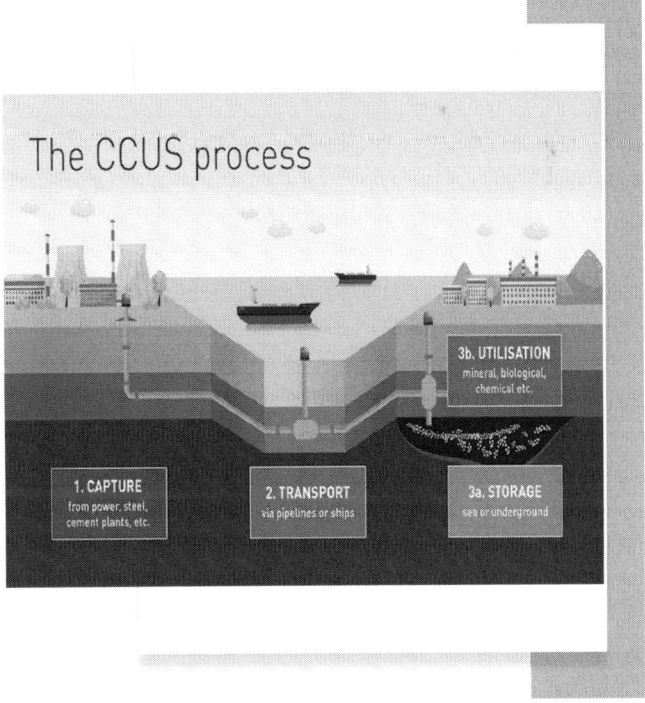

CCUS?

"10억 톤 규모의 CCUS 기술을 가진 분에게는
1억 달러를 드리겠습니다."

- ➤ XPRIZE Carbon Removal Project,
  '21년 1월 Tesla의 Elon Musk가 내건 상금

- ➤ '25년 4월22일(지구의 날) 우승자 발표
  (1천 만톤/년의 CO_2를 100년 이상 격리할 수 있는 기술)

Source : NTIS News (2010210)

CCUS?

- ➤ **CCUS : CCS + CCU**
  - ✓ CCS : Carbon Capture and Storage
  - ✓ CCU : Carbon Capture and Utilization

- ❖ CCUS : 탄소를 포집, 저장하는 것뿐만이 아닌 활용을 통해 부가가치가 높은 유용자원 또는 물질로 전환하는 기술

Source : GS Caltex Newsletter (2023)

CO₂의 활용

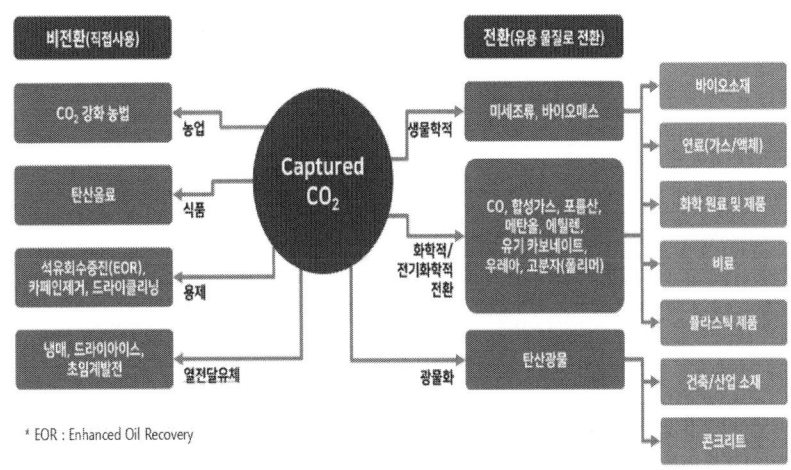

Source : 한국에너지기술연구원, CCUS 심층투자 분석보고서 (2021)

참고) CO₂ to Chemical

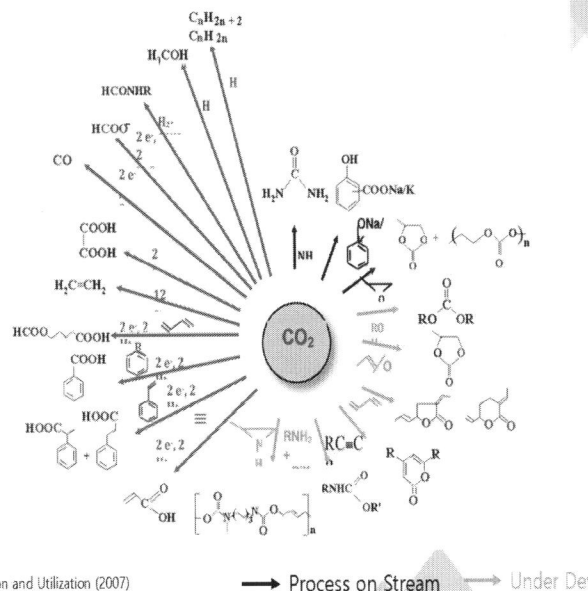

Source : GHG Mitigation and Utilization (2007)

CCUS Concept

Source : 탄소중립을 위한 CCUS 기술동향 (2021)

Carbon Capture Overview

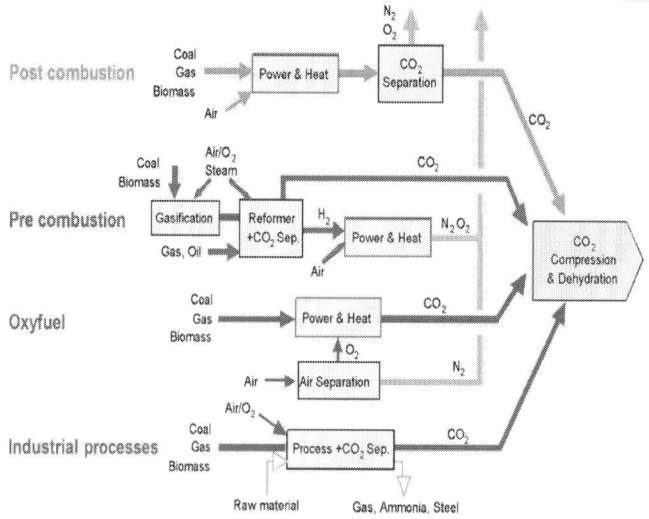

Overview of CO₂ Capture Processes and Systems

Source : IPCC (2005)

참고) Post combustion vs Pre combustion

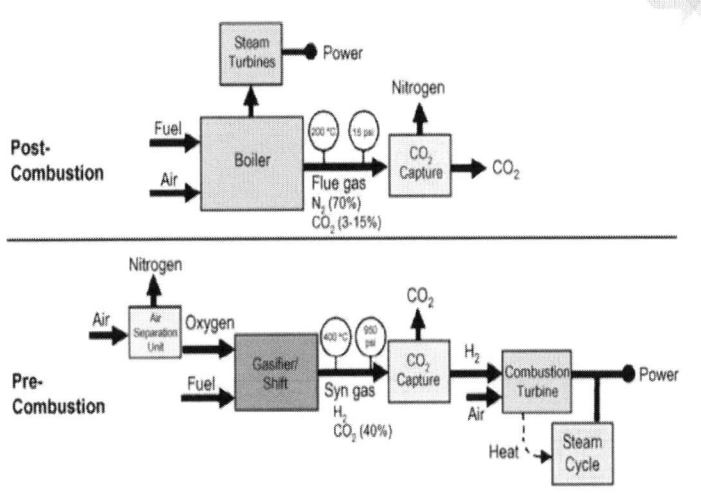

Source : 2018 2nd International Conference on Advances in Energy, Environment and Chemical Science, Advanced in Engineering Research vol 155.

연소 후 CO_2 포집기술

➢ 연소 후 CO_2 포집기술 비교

구분	종류	장점	단점
흡수분리	습식	· 대용량 가스 처리에 용이 · 이산화탄소 농도변화에 적용성이 큼	· 흡수제 재생에 다량의 에너지 소비 · 흡수제 열화 및 재료부식
	건식	· 저농도 대용량 가스분리 가능 · 고온·고압의 가스시스템에 적용가능	· 장치 및 운전이 복잡 · 기-고 반응으로 반응속도가 느림
흡착분리		· 장치와 운전이 비교적 간단 · 환경영향 및 에너지 효율 우수	· 비정상상태에서의 운전(분리효율 낮음) · 대용량 처리 곤란 및 흡착제 비활성화
막분리		· 장치와 운전이 비교적 간단 · 에너지 소비가 적음	· 대용량화 곤란(모듈 복합체 고가 시설비) · 분리막의 열화로 내구성 취약
증류분리(심냉법)		· 투자비가 저렴 · 오랜 경험으로 공정의 신뢰도가 높음	· 에너지 소비가 많음 · 대용량 가스처리에 곤란

❖ CCS 전체 처리비 중, CO_2 포집 부분이 70~80% 비중으로 비용 절감 필요

Source : 연소 후 이산화탄소 포집기술 현황 (2019)

CO_2 포집기술 동향

Source : 이산화탄소 포집기술 국내외 기술 동향 (2020)

구분		2세대 기술	3세대 기술
연소후 포집	습식 포집 기술	· Precipitating solvents · Two phase liquid system · Ionic fluids	· Precipitating solvents · Two phase liquid system · Enzymes · Ionic fluids · Encapsulated solvents · Electrochemical solvents
	건식 포집 기술	· Calcium looping systems	· Other looping systems · Vaccum pressure swing (VPS) · Temperature Swing (TS)
	분리막 기술	· Polymeric membranes · Polymeric w/cryogenic	
	기타	· Cryogenic · CO_2 enriched flue gas · Pressurized post-combustion	· Cryogenic · Supersonic · Hydrates · Algae · Pressurized post-combustion · MCFC
연소전 포집	건식 포집 기술	· Sorption Enhanced Water Gas Shift (SEWGS) · Sorption Enhanced Steam-Methane Reforming (SE-SMR)	· Sorption Enhanced Steam-Methane Reforming (SE-SMR)
	분리막 기술	· Metal and composite membranes · Ceramic membranes	· Metal and composite membranes · Ceramic membranes
	기타	· Concepts with fuel cells	· Cryogenic · Concepts with fuel cells
순산소 포집			· Chemical Looping Combustion · Oxygen transporting membranes(OTM) · Pressurized oxy-combustion

Carbon Capture : 한국전력 활용 사례

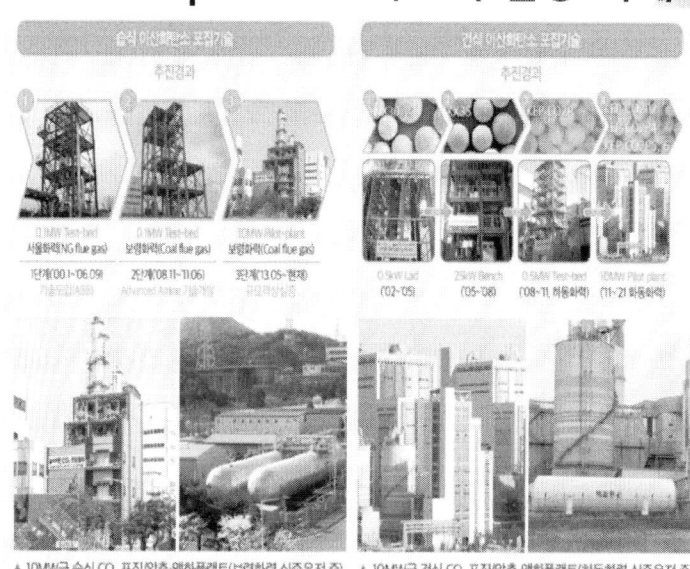

▲ 10MW급 습식 CO_2 포집/압축·액화플랜트(보령화력 실증운전 중) ▲ 10MW급 건식 CO_2 포집/압축·액화플랜트(하동화력 실증운전 중)

Source : 전력연구원 자료 (2021)

Carbon Capture : Pilot & Plant, 한국 (1/2)

기술	분리소재	공정특징	적용 분야	규모	성능 지표	성능 수치	수행기관
습식 공정	아민계	KOSOL	석탄화력	10MW (180t-CO_2/d)	생산 CO_2 순도(%)	99	전력연(특허) DL E&C(실시)
					재생열(GJ/t-CO_2)	2.5-2.6	
		MAB	석탄화력	0.5MW (10t-CO_2/d)	생산 CO_2 순도(%)	99	에너지연(운전) 서강대(공정) 경희대(소재) KCRC(특허)
					재생열(GJ/t-CO_2)	2.2-2.4	
	K_2CO_3 +아민계	KIERSOL	석탄화력, 시멘트	0.5MW (10t-CO_2/d)	생산 CO_2 순도(%)	99	에너지연(특허) SCT Eng.(실시) SK MR(실시)
					재생열(GJ/t-CO_2)	2.2-2.4	
	암모니아수	가압 재생	석탄발전 배가스	0.03MW (100Nm^3/h)	생산 CO_2 순도(%)	99.9 @6.5bar	에너지연
					재생열(GJ/t-CO_2)	2.7	
		상압 재생	제철소(BFG)	0.5MW (10t-CO_2/d)	생산 CO_2 순도(%)	99	포항산업과학원
					재생열(GJ/t-CO_2)	2.2-2.4	

Source : 한국에너지기술연구원, CCUS 심층투자 분석보고서 (2021)

Carbon Capture : Pilot & Plant, 한국 (2/2)

기술	분리소재	공정특징	적용 분야	규모	성능 지표	성능 수치	수행기관
건식 공정	K_2CO_3 기반	유동층 공정	석탄화력	10MW (150t-CO_2/d)	생산 CO_2 순도(%)	99.9	에너지연(공정) 전력연(소재)
					CO_2 제거율(%)	80	
					재생열(GJ/t-CO_2)	4.4	
	제올라이트 기반	VSA공정	석탄발전 배가스	110Nm^3/h	생산 CO_2 순도(%)	99	에너지연
					CO_2 제거율(%)	90	
					전력소비량(kWh/Nm^3 CO_2)	0.6	
막분리	고분자막	4단 분리막	바이오가스 정제	100Nm^3/h	생산 CO_2 순도(%)	95	화학연
					메탄회수율(%)	98	
순산소연소	초임계 순산소연소	연소중포집 (후처리설비 불필요)	가스터빈 발전	0.5MW (35t-CO_2/d)	생산 CO_2 순도(%)	98	에너지연
					CO_2 제거율(%)	99	
					NOx, SOx 배출농도(ppm)	10 이하	
	순환유동층 순산소연소	연소중포집	석탄화력발전	0.1MWth (10MWth 운전 중)	생산 CO_2 순도(%)	97	에너지연
					NOx, SOx 배출농도(ppm)	10 이하	
매체순환연소	산소전달 입자	유동층 공정	LNG 발전 매립지가스, 바이오가스	0.5MWth 실증 완료 3MWth 건설 중	생산 CO_2 순도(%)	98	에너지연 전력연
					NOx 배출농도(ppm)	15 이하	

Source : 한국에너지기술연구원, CCUS 심층투자 분석보고서 (2021)

Carbon Capture : DAC 기술

> DAC : Direct Air Carbon Capture

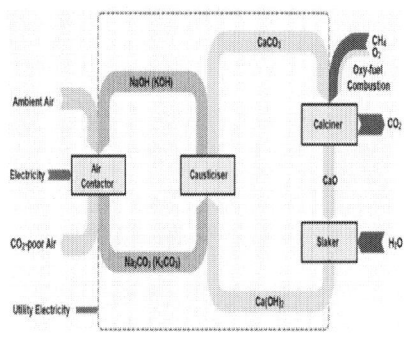
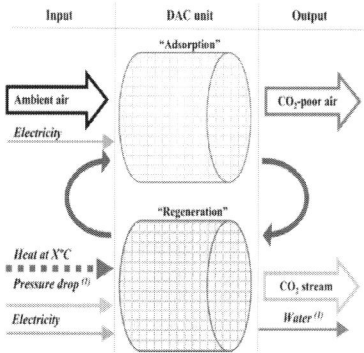

고온 DAC 기술
(using NaOH or KOH solution, @ 900℃)

저온 DAC 기술
(using solid sorbent, @100℃)

Source : Jr. Clean Product 224 (2019)

Carbon Storage Concept

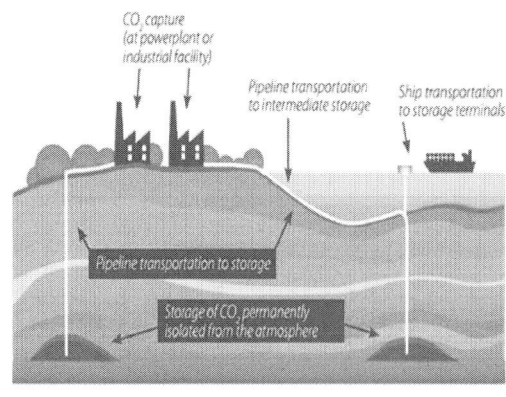
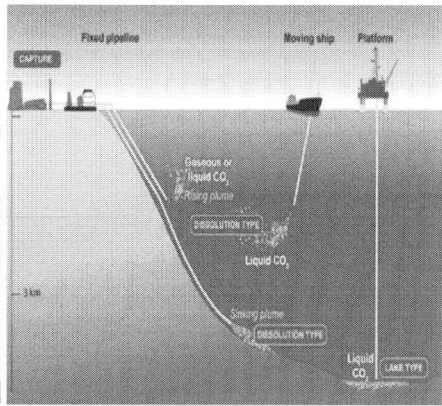

Source : Carbon Capture and Storage Program (2016), IPCC (2005)

주요 국가 CCUS 목표

> 주요 국가별 GHG 감축 목표 및 CCUS 기술 기여도

구분	기술	목표년도	기준년도	감축량	CCS 또는 CCUS 기술 비중
미국	CCUS	2032	2005	-	9%
영국	CCUS	2032	2018	40 MtCO₂	9%
프랑스	CCS	2050	1990	-	2.5%
중국	CCS	2060	2050 (BAU)	519 MtCO₂	8.8%
한국	CCUS	2050	2018	85~95 MtCO₂	11.7~13.1%

* BAU : Business as Usual (배출 전망치)

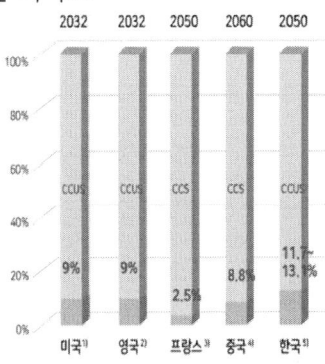

1) 미국 The White House(2016), United States Mid-Century Strategy for Deep Decarbonization
2) 영국 정부(HM Government)(2020), The Ten Point Plan for a Green Industrial Revolution
3) 프랑스 정부(2020), National Low Carbon Strategy
4) 중국 Energy Foundation China(2020), China's New Growth Pathway : From the 14th Five-Year Plan to Carbon Neutrality
5) 환경부(2021), 탄소중립 시나리오 초안

Source : 한국에너지기술연구원, CCUS 심층투자 분석보고서 (2021)

CCS Cost Estimation

> CCS Cost (IEA, 2010)

	Organization, region				
	MIT, USA	NETL[1], USA	GCCSI[2], Australia	IEA-GHG[3], UK	NZEC[4], China
Year of cost data	2007	2007	2009	2009	2009
Cost of CO₂ avoided (USD/tCO₂)	58	69	74	59	42

[1] NETL: National Energy Technology Laboratory, [2] GCCSI: Global Carbon Capture & Storage Institute, [3] IEA-GHG: International Energy Agency Greenhouse Gas R&D Programme, [4] NZEC: Near Zero Emission Control project.

Source : Jr. Climate Change 7(2) (2016)

강 의 노 트

🚗 1.

🚗 2.

🚗 3.

제5장
화학공장 Process 설계 1

탄소중립 관련 추가 내용

RE100 개요

❖ 기업이 필요로 하는 전력의 100%를 태양광, 풍력 등 친환경 재생에너지로 사용하겠다고 선언하는 자발적인 글로벌 기업 리더십 이니셔티브

RE100 vs CF100

RE100과 CF100 비교

RE100	구분	CF100
재생에너지 100% 사용	의미	무탄소 에너지원 100% 사용
풍력·태양광 등 재생에너지	주요에너지원	재생에너지+원자력+ 수소연료전지
더 클라이밋 그룹 다국적 비영리기구 (2014년)	발족	구글·UN에너지· 지속가능에너지기구 등 (2018년)
국내 31개 포함 글로벌 407개 기업	참여기업수	글로벌 119개 기업

Source : 2023.06.06 Biz-watch

RE100 대신 원전포함 CF 100을...

➢ 한국전력 정산 단가 (kWh당 도매가격, '22년 기준)

- 원전 : 52원

- LNG : 239원

- 신재생 에너지 : 271원

➢ 한국전력 판매 단가 : 109원

→ 재생에너지 증가할수록 적자 증가

→ 전력망에서 재생에너지 20% 넘으면 전력 품질 유지 불가

➢ Google이 RE100을 달성했다는 것에는 원자력을 이용한 무탄소 전기도 포함

Source : 2023.09.21 한겨레

초소형 원자로의 등장

➢ 미국 INL (Idaho National Lab.)

SMR 10분의 1 크기 '마이크로 원자로' 5000명 쓸 전력 공급

건설 아닌 찍어내는 원전...쉽게 옮기고 섬·오지에도 설치

* SMR : Small Modular Reactor (소형 모듈 원전)

Source : 2023.10.03 한국경제

신임 IPCC 의장 발언

➢ 2050년 탄소중립 되어도 기후위기 악화될 수도...

- 계속 늘어나는 CO_2 누적 배출량

- 석유 매장량의 30%,
  가스 매장량의 50%,
  석탄 매장량의 80%가 땅에 남아야 함

- 각종 조치를 연기하는 경로를 택할 경우, 탄소중립 달성해도 지구온도는 2℃보다 더 상승할 수 있다고 경고

Source : 2023.10.02 영국 Guardian Interview 내용

들어가기 앞서

집 짓기 절차

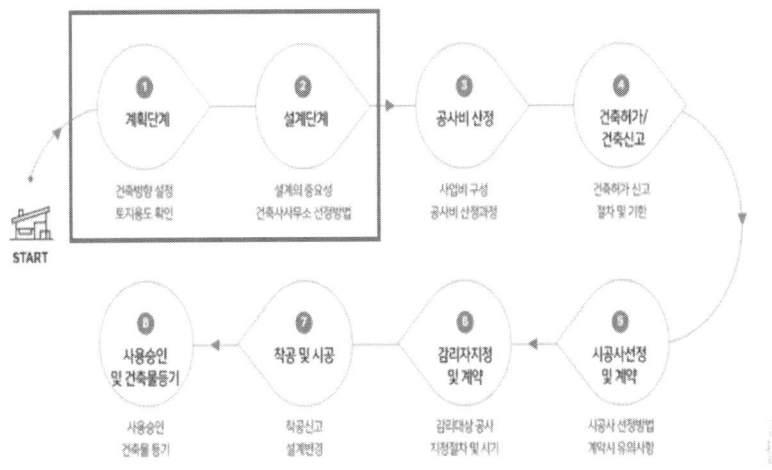

Source : 대한건설협회

화학공장은?

◆ 제5장. 화학공장 Process 설계 1 ▶ 109

화학 관련 공정을 위한 전형적인 설계 과정

Table 1-1 Typical design steps for chemical and biochemical processes

1. Recognize a societal or engineering need.
   a. Make a market analysis if a new product will result.
2. Create one or more potential solutions to meet this need.
   a. Make a literature survey and patent search.
   b. Identify the preliminary data required.
3. Undertake preliminary process synthesis of these solutions.
   a. Determine reactions, separations, and possible operating conditions.
   b. Recognize environmental, safety, and health concerns.
4. Assess profitability of preliminary process or processes (if negative, reject process and create new alternatives).
5. Refine required design data.
   a. Establish property data with appropriate software.
   b. Verify experimentally, if necessary, key unknowns in the process.
6. Prepare detailed engineering design.
   a. Develop base case (if economic comparison is required).
   b. Prepare process flowsheet.
   c. Integrate and optimize process.
   d. Check process controllability.
   e. Size equipment.
   f. Estimate capital cost.
7. Reassess the economic viability of process (if negative, either modify process or investigate other process alternatives).
8. Review the process again for environmental, safety, and health effects.
9. Provide a written process design report.
10. Complete the final engineering design.
    a. Determine equipment layout and specifications.
    b. Develop piping and instrumentation diagrams.
    c. Prepare bids for the equipment or the process plant.
11. Procure equipment (if work is done in-house).
12. Provide assistance (if requested) in the construction phase.
13. Assist with start-up and shakedown runs.
14. Initiate production.

Source : Plant Design & Economics for Chemical Engineers (5th ed.)

- 정유 공장 건설 사례 동영상

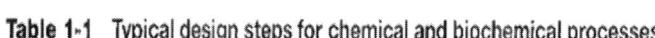

Table 1-1 Typical design steps for chemical and biochemical processes

1. Recognize a societal or engineering need.
   a. Make a market analysis if a new product will result.
2. Create one or more potential solutions to meet this need.
   a. Make a literature survey and patent search.
   b. Identify the preliminary data required.
3. Undertake preliminary process synthesis of these solutions.
   a. Determine reactions, separations, and possible operating conditions.
   b. Recognize environmental, safety, and health concerns.

4. Assess profitability of preliminary process or processes (if negative, reject process and create new alternatives).
5. Refine required design data.
   a. Establish property data with appropriate software.
   b. Verify experimentally, if necessary, key unknowns in the process.
6. Prepare detailed engineering design.
   a. Develop base case (if economic comparison is required).
   b. Prepare process flwsheet.
   c. Integrate and optimize process.
   d. Check process controllability.
   e. Size equipment.
   f. Estimate capital cost.
7. Reassess the economic viability of process (if negative, either modify process or investigate other process alternatives).
8. Review the process again for environmental, safety, and health effects.
9. Provide a written process design report.
10. Complete the final engineering design.
    a. Determine equipment layout and specifications.
    b. Develop piping and instrumentation diagrams.
    c. Prepare bids for the equipment or the process plant.
11. Procure equipment (if work is done in-house).
12. Provide assistance (if requested) in the construction phase.
13. Assist with start-up and shakedown runs.
14. Initiate production.

화학 공장 설계 Process (Block Diagram)

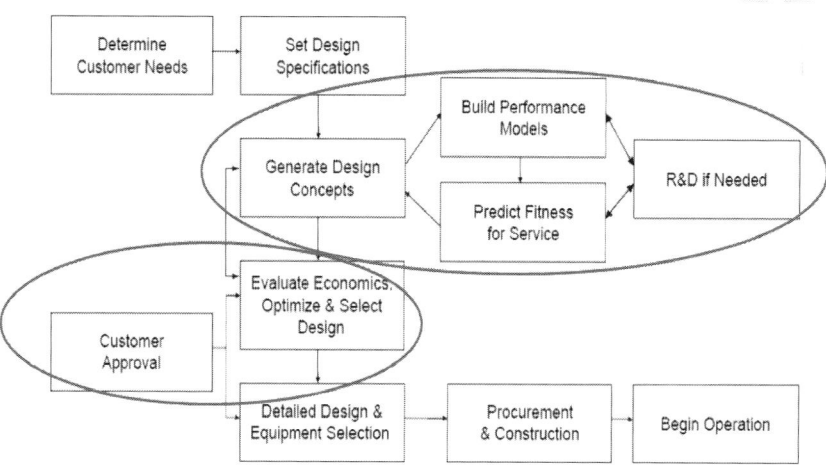

화학 공장 설계 제약/고려사항

Source : Chemical Plant Design

화학공정설계 vs 화학제품 설계

| PROCESS DESIGN | PRODUCT DESIGN |

화학제품 설계 전략 (1/3)

1) **Need**
   - A need for the product must be identified
   - In the case of a commodity chemical
     - Once there is a market demand for the product, a new plant is constructed to produce it, probably using the same technology
   - In chemical product design
     - Once the need is established, the search for the best product begins
   - To recognize / identify a need
     - Chemical companies devoted to product design deal with customers all the times
     - Customers are interviewed, results are interpreted, need defined
     - Sometimes even the customers do not realize the need, but when the product is introduced to them, they like it.

2) **Ideas**
   - Search for the best product begins
   - Just as in brainstorming, when ideas are generated, there are no bad ideas
   - The chances of the first idea being generated being the best are slim or none
   - As many ideas as can be imagined should be generated before moving on to the selection step

Source : Chemical Plant Design

화학제품 설계 전략 (2/3)

3) **Selection**
   - It is necessary to screen the ideas and select a few more detailed investigation
   - Scientific principles can be applied
   - Eliminate
     - If thermodynamically impossible
     - If unfavorable kinetics (or downgrade)
     - If expensive (or downgrade)
   - Don't reject too soon if in doubt
   - Method
     - Concept screening
       - ✓ A selection matrix prepared by listing criteria used to evaluate alternatives
     - Concept scoring
       - ✓ Only on alternatives survived the concept screening process
       - ✓ Based on weightage to product specification according to their importance to the function of the product and customer preference

Source : Chemical Plant Design

화학제품 설계 전략 (3/3)

4) **Manufacture**
   - Determine whether the product can be manufactured
   - Developing detailed product specifications
   - Determining how the product is to be manufactured
   - Estimating the cost of manufacturing
   - Sample or prototype testing

Source : Chemical Plant Design

요약해 보면....

> Product & Market

> Best Process

> Profitability & Economics

> SHE : Safety, Health & Environment

MSDS

> MSDS (Material Safety Data Sheet)

: 화학물질의 유해, 위험성, 취급방법, 응급조치요령 등을 기술하는 자료로 화학물질을 안전하게 사용하기 위한 설명서

HAZOP

➢ HAZOP (Hazard and Operability)

: 모든 공정이나 장치의 위험물과 조업에 문제점이나 위험성을 찾아내는 체계적인 조사

❖ 공정의 모든 부분 (Pipeline, 장비, 설비 및 계장 등)의 정상 조건에서 벗어날 수 있는 모든 가능성을 찾아내어 안전조업이 가능할 수 있도록 함

Safety Issue

- 중국 화학공장 폭발 사고

- 미국 비료공장 폭발 사고

- 베이루트 폭발 사고

- 미국 Chocolate 공장 폭발 사고

폭발의 3요소

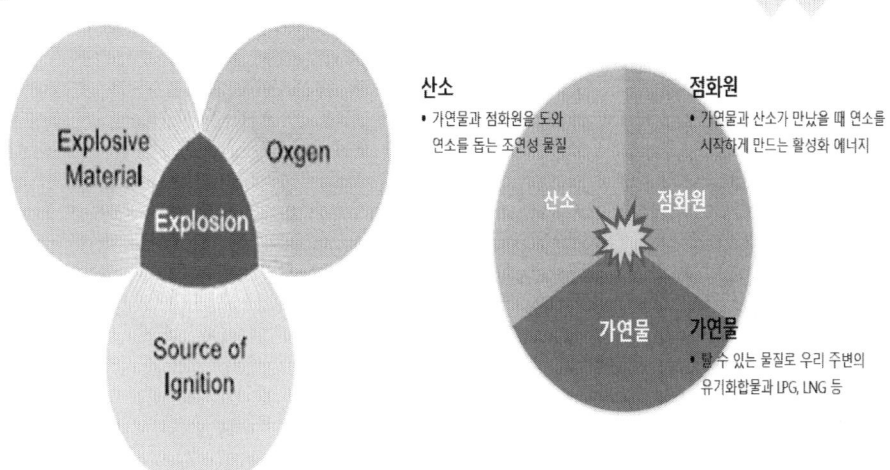

산소
- 가연물과 점화원을 도와 연소를 돕는 조연성 물질

점화원
- 가연물과 산소가 만났을 때 연소를 시작하게 만드는 활성화 에너지

가연물
- 탈 수 있는 물질로 우리 주변의 유기화합물과 LPG, LNG 등

Flammable Range

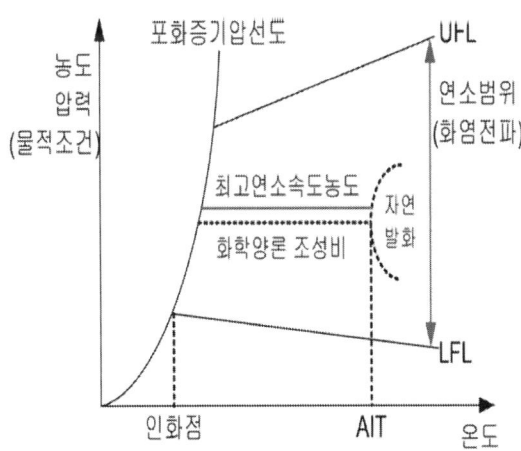

* UFL: Upper flammable limit, LFL : Lower flammable limit
* AIT : Auto ignition temperature

Propylene Flammable Range

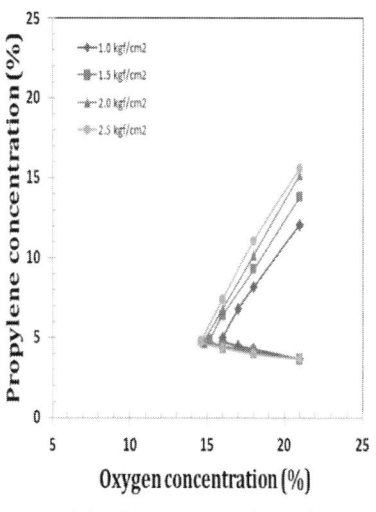

Relation between O₂ and propylene concentration (@200°C)

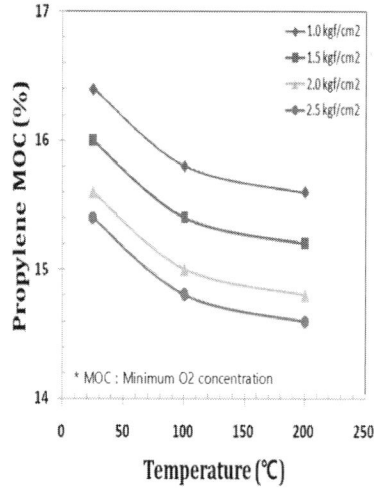

Relation between O₂ concentration and temperature

Health & Environment Issue (1/2)

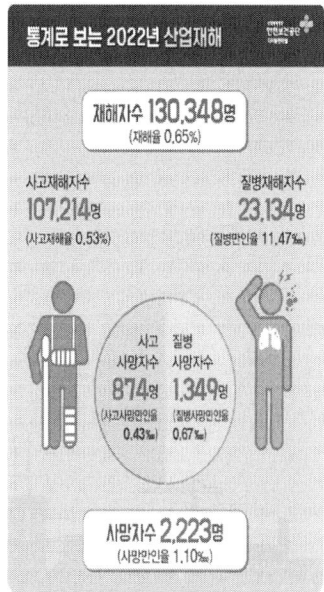

벤젠 중독 위험

- 백혈병, 빈혈을 일으킴
- 취급 시 환기 실시 및 보호구 착용
- 건강이상 시 의사와 상담

❖ 증상 : 현기증, 건강검진에서 빈혈이나 백혈구 이상 등

Health & Environment Issue (2/2)

➢ 연소배기 가스 및 유해물질

➢ 공정 폐수 및 폐기물

➢ 소음, 진동, 악취

➢ 기타 : 지반침하, 방사능 등

강 의 노 트

1.

2.

3.

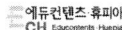

제6장
화학공장 Process 설계 2

Brief review from Lab to Plant Design

Product Concept

Example : Acrylic acid

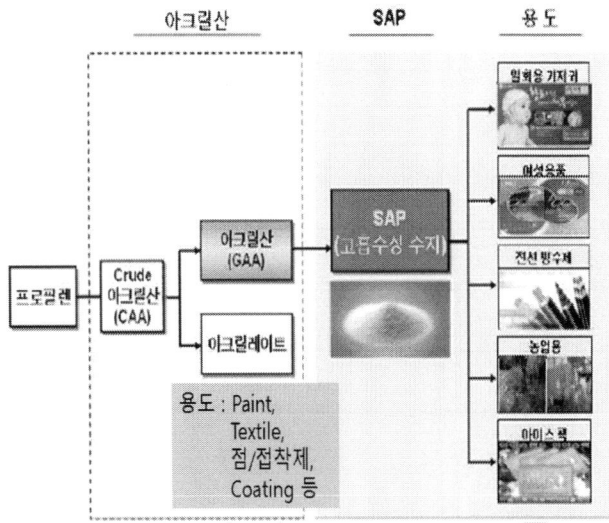

성균관대 석유화학 강의 자료

SAP : Super Absorbent Polymer

시장 전망 Survey (1/3)

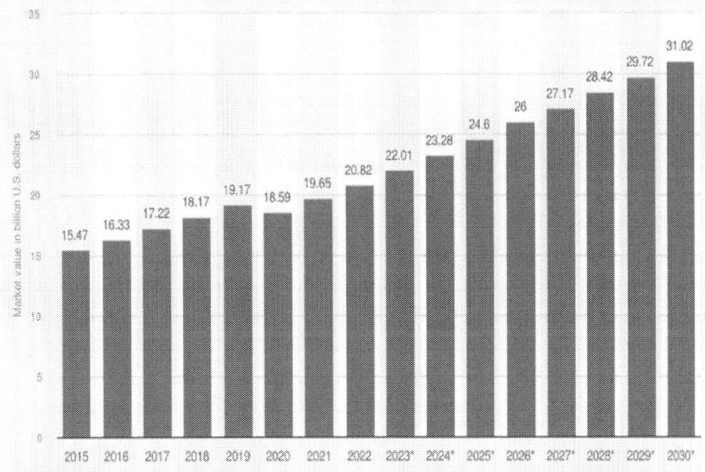

Source : AgileIntel Research(2023)

시장 전망 Survey (2/3)

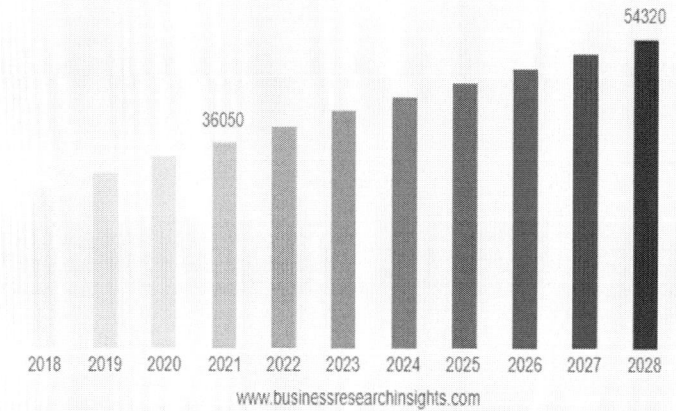

시장 전망 Survey (3/3)

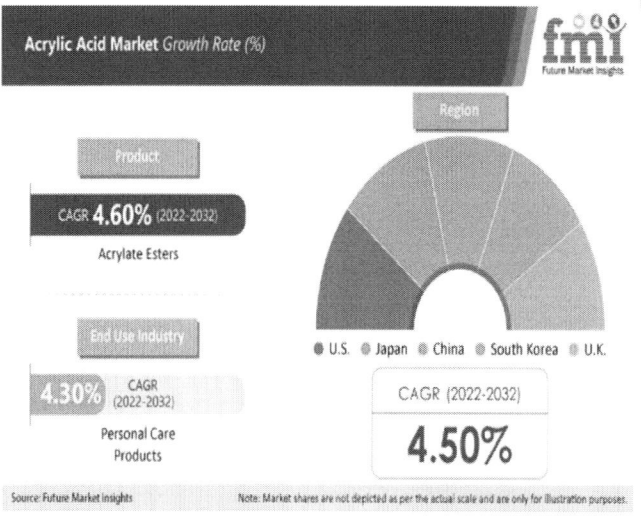

CAGR : Compound Annual Growth Rate

참고) CAGR

> CAGR : 연평균 성장률로 표현 (연평균 복합 성장률)
> → 시간에 따라 일정한 비율의 변화를 제공하는 비즈니스, 투자용어

$$\text{CAGR}(t_0, t_n) = \left(\frac{V(t_n)}{V(t_0)}\right)^{\frac{1}{t_n - t_0}} - 1$$

$V(t_0)$는 시작값, $V(t_n)$는 끝값, $t_n - t_0$는 년수

년도	매출액(억 원)	전년대비 증가(%)
2015	100	
2016	150	50.0
2017	250	66.7
2018	400	60.0
2019	450	12.5
2020	500	11.1

산출평균 증가(%) : ?
CAGR(%) : ?

시장/기술 동향 기준 주요 자료

"IHS Markit Report" (Former, "Chemical Economics Handbook (CEH)")

"Process Economics Program (PEP) - Yearbook"

"ChemLOCUS"

"화학경제연구원"

"HS (Harmonized System) Code"

"www.kitco.com"

참고) IHS자료 포함 주요 사항

- New Executive Summary: Key commercial implications of the developments contained in the report
- Summary: Global supply-demand for individual chemicals or groups of chemicals, future growth with five-year projections and historical data
- Demand: Analysis and forecasts, including market size, end-use applications, consumption trends and competing materials
- Supply: Producers, plant locations, annual capacities, capacity utilization, production volumes and trends
- Process: Commercial manufacturing processes involved and basic chemistry used
- Prices: Histories, annual unit sale volumes and factors affecting prices
- Trade: Import-export data, including countries of origin and destination and shipment values

참고) PEP Yearbook 포함 주요 사항

- Production economic data for more than 1,500 processes used to manufacture over 600 chemical, polymer, refining and biotech products
- Estimates for raw material and utility requirements
- Capital and production costs for three plant capacity levels
- Customization of plant capacity for quick scaling analysis
- Cost information for the US Gulf Coast, Germany, Japan and China, with limited process coverage for Canada and Saudi Arabia
- Conversion of data to either English or metric units

참고) HS Code

➢ HS : 1988년 국제협약으로 채택된 국제통일상품분류체계의 약칭
(Harmonized Commodity Description and Coding System)

❖ 국제협약 HS code

- 1~2 자리 : 상품 군별 분류
- 3~4 자리 : 소분류, 동일류 내 품목의 종류/가공도별 분류
- 5~6 자리 : 세분류, 동일호 내 품목의 용도/기능에 따른 분류
- ✓ 7자리부터는 각 나라에서 세분하여 부여
- ✓ 한국은 10자리를 사용

➢ 품목별 수출입 통계 및 관세율 확인

참고) HS Code

참고) 다양한 자료, 협력기관 등 조사

"Patent"

"기존 계약서 면밀 검토"

"Collaboration Company"

"Co-work 외부기관 또는 학교"

"기술 도입 from Licensor"

"Joint Venture"

- 기존 기술
- 신규 기술
- 독점적 기술
- 기술보호 가능성

What is Priority No.1 Consideration?

Acrylic acid 제조 방법 (문헌 조사)

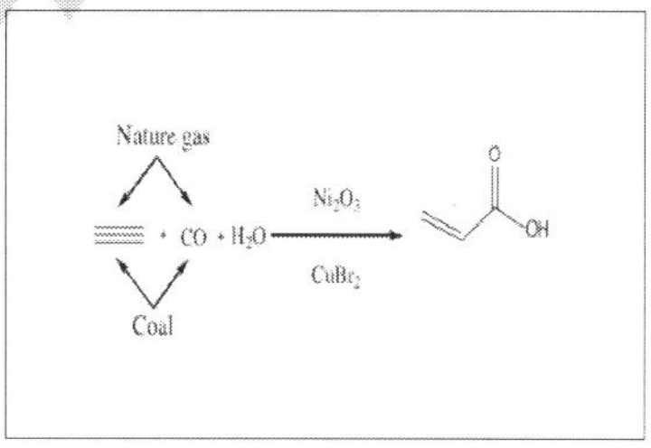

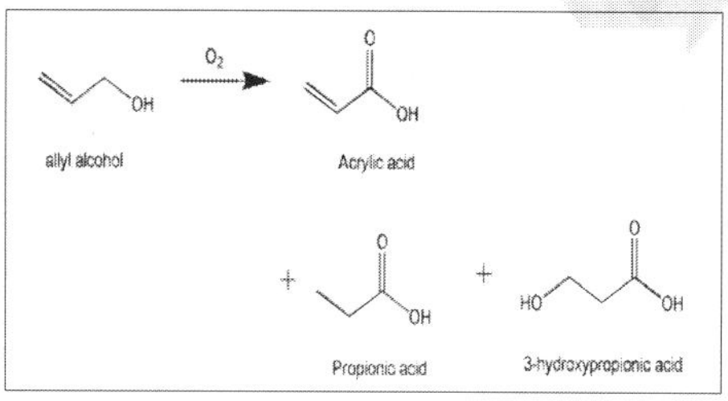

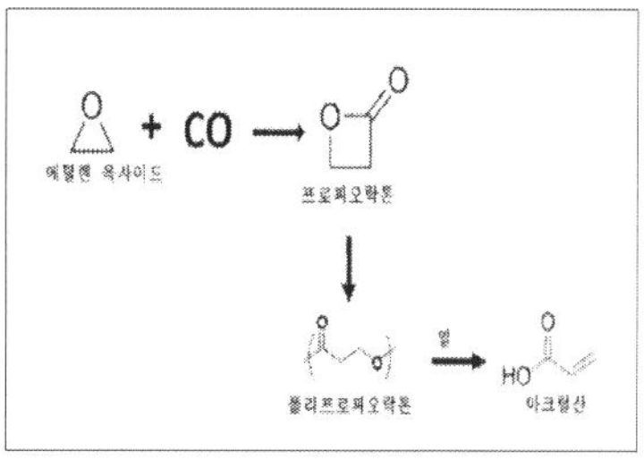

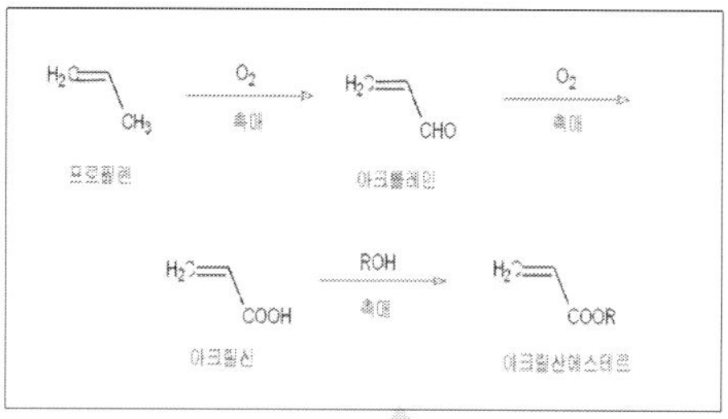

Acrylic acid 제조 방법 (문헌조사)

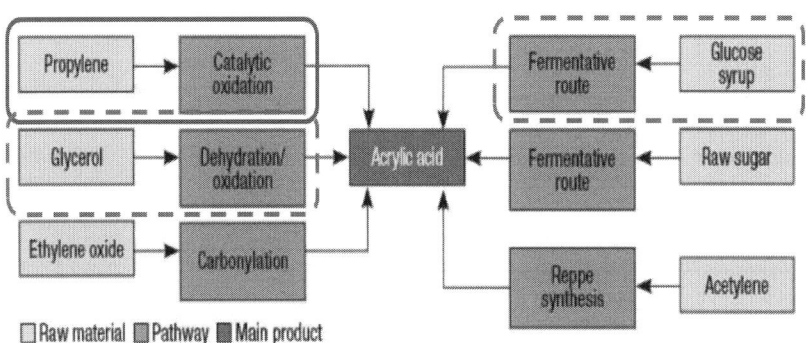

☐ Raw material ☐ Pathway ■ Main product

What is the best for us now?
And then?

Source : Chemical Engineering Feb 2016

View for Research vs or & Business

"Key or Core Technology"

"New or Creative Output"

"Technology Level"

"Plausible Process"

etc.......

"Time to Market"

"Capital Investment"

"Operability"

"Profitability"

etc.......

화학제품 연구개발 및 상업화 Procedure

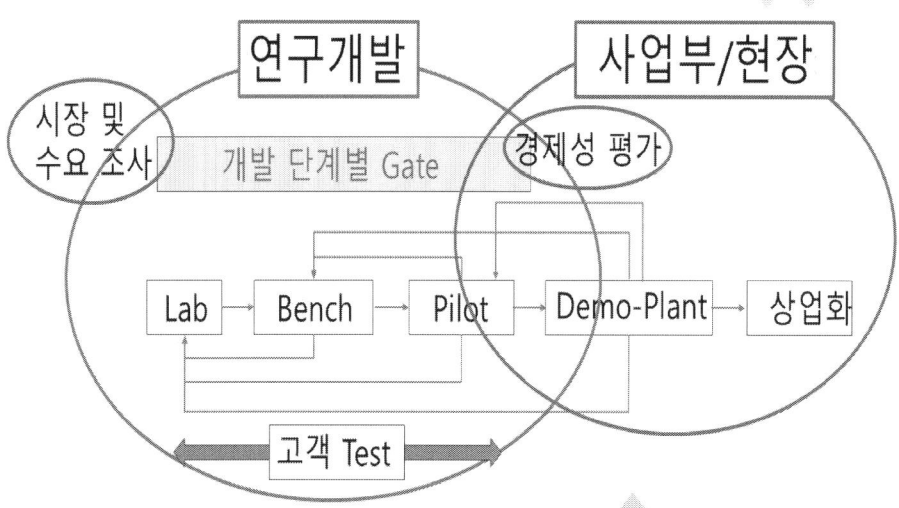

Acrylic acid (Conceptual process)

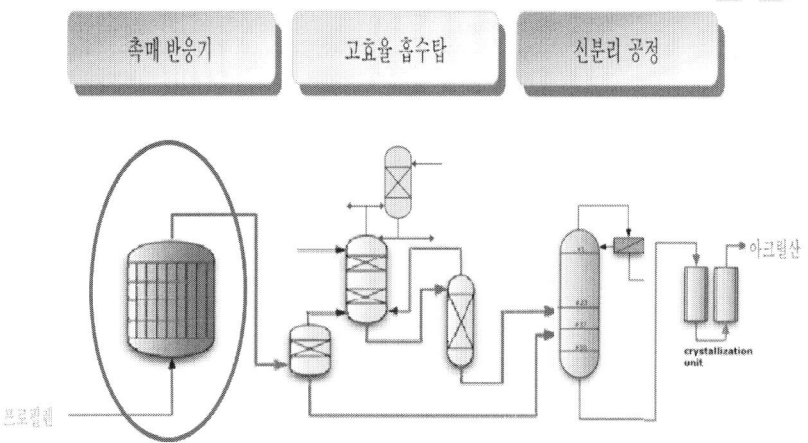

What is the key development area?

반응공정 개발을 위한 주요 요소

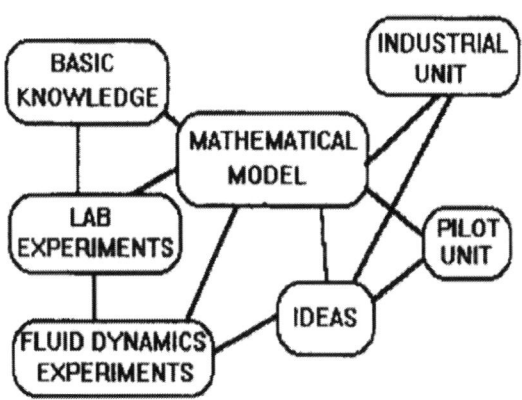

Source : Catalysis Today 34 (1997)

대표적인 화학공정 반응기 형태

Continuously Stirred Tank Reactor (CSTR) – complete mixing (uniform concentration everywhere)

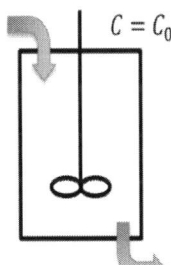

$C = C_0$

Plug Flow Reactor (PFR) – zero axial mixing (spatially varying concentration)

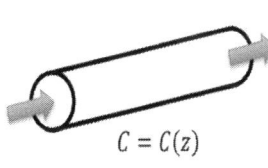

$C = C(z)$

CSTR scale-up 대표적 고려 사항 (Mixing)

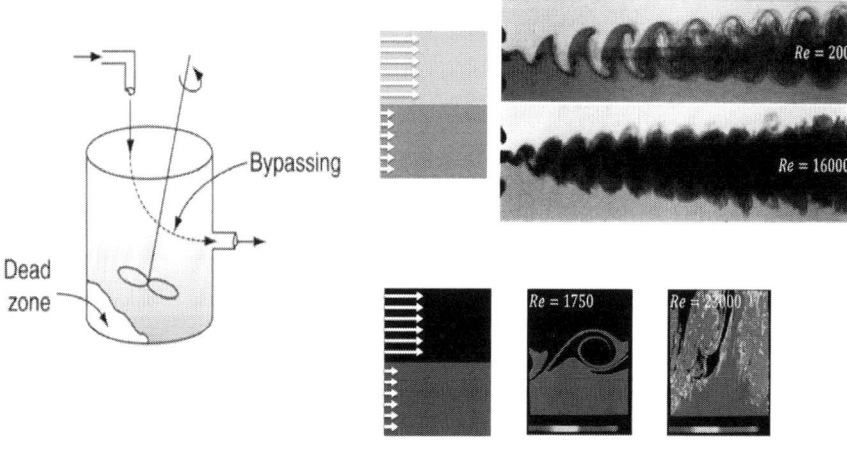

CSTR scale-up

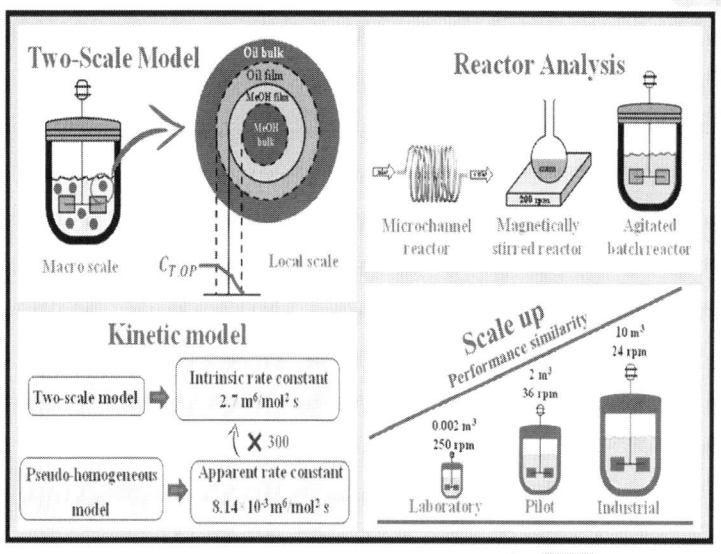

Tubular Reactors (Lab 촉매 반응기)

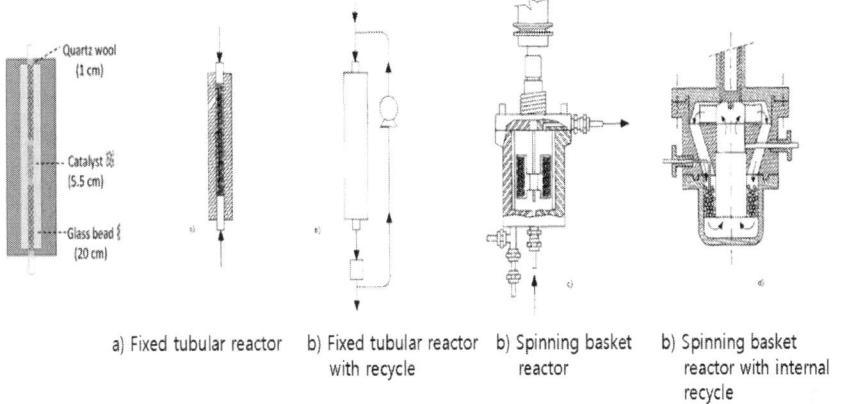

a) Fixed tubular reactor
b) Fixed tubular reactor with recycle
b) Spinning basket reactor
b) Spinning basket reactor with internal recycle

Source : Catalysis Today 34 (1997)

Tubular Reactor Scale-up (촉매 반응)

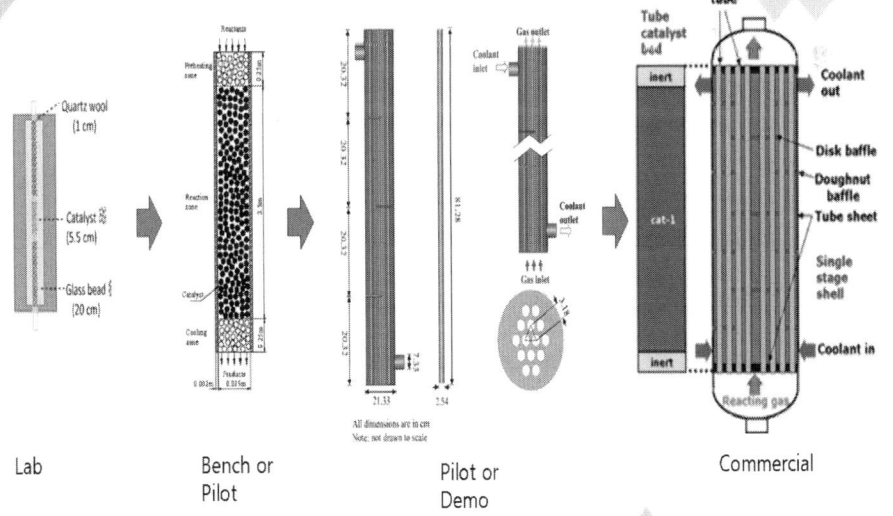

Lab   Bench or Pilot   Pilot or Demo   Commercial

Tubular Reactor Scale-up Issue (촉매 반응)

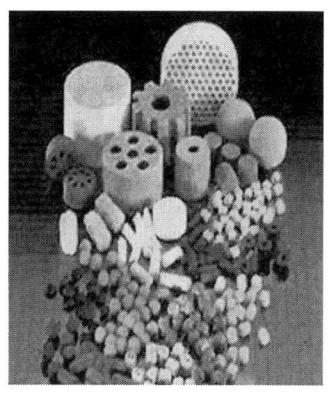

촉매의 다양한 형상

Source : Catalyst Handbook

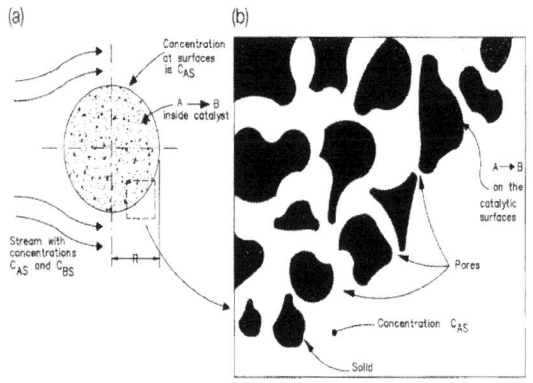

기공이 있는 구형 촉매에서의 Diffusion

Source : Catalysis Today 34 (1997)

참고) 다양한 형태의 촉매 반응기

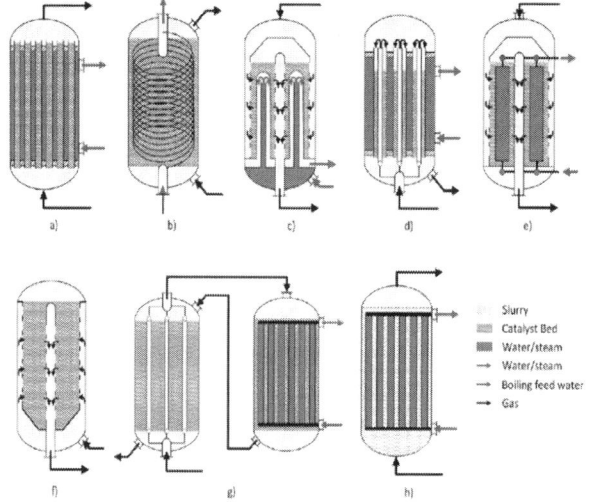

Simplified reactor layout of (a) Lurgi tubular reactor, (b) Linde Variobar, (c) Toyo MRF, (d) Mitsubishi Superconverter, (e) Methanol Casale IMC, (f) Haldor Topsøe adiabatic reactor, (g) Lurgi MegaMethanol and (h) Air Products LPMEOH, adapted from Buttler 2018. 138

Acrylic acid Process

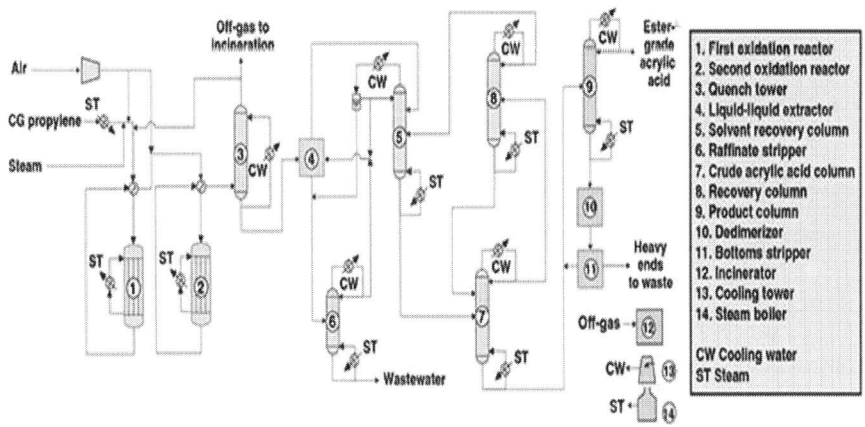

❖ Material, Heat & Utility Balance ❖ Equipment material, sizing
❖ Process operation & Safety ❖ Analysis method

Source : Chemical Engineering Feb 2016

사업타당성 분석

시장성 분석	기술성 분석	재무 분석	경제성 분석
제품 개발 관련 정보 수집 시장 특성 분석 시장규모 분석 소비자 분석 미래수요예측 제품 경쟁력 분석 마케팅전략	기술의 타당성 생산시설의 입지 생산자원, 생산능력 시설계획 생산방법 및 공정	추정 재무상태표 (자산, 자본, 부채 추정), 추정손익계산서 (매출액, 매출원가, 판매비 및 관리비, 영업이익 추정)	수익전망 투자수익률 손익분기점 분석

> 종합하여 투자 진행여부 결정

아래 제품에 대한 사업타당성은?

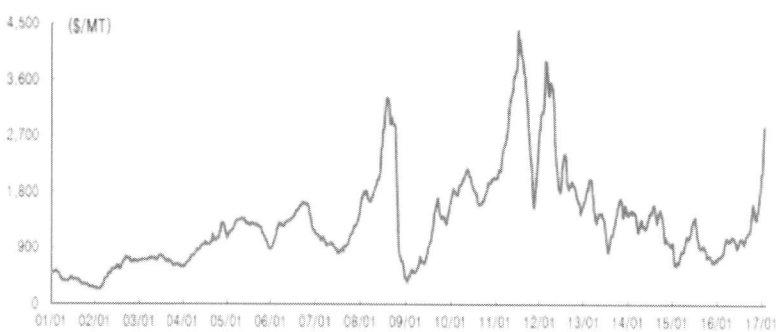

Go or Stop

공장건설 Project의 Matrix 조직 (예)

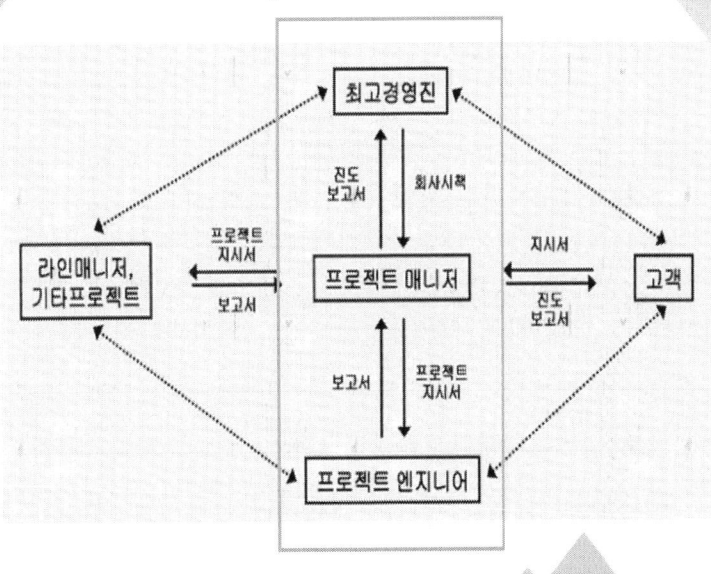

Plant Layout

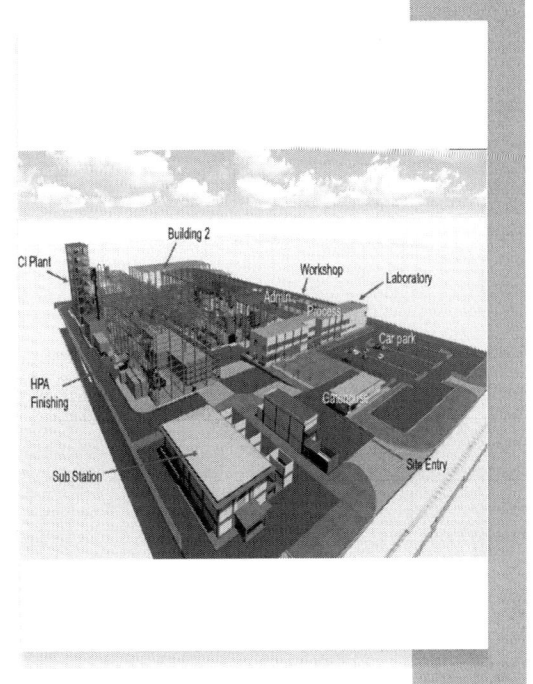

화학공장건설을 위한 필수 자료

> Basic Design Package
  - Design Basis
  - PFD (Process Flow Diagram) & UFD (Utility Flow Diagram)
  - P&ID (Piping & Instrumentation Diagram)
  - Equipment List (Material, T, P, Capacity etc.)
  - Operation Manual (w/ Start-up, Shut-down, Emergency)
  - 기타 : 분석 방법, MSDS, 물성 정보 등

> Detail Engineering
  - Plot Plan & Utilities
  - Equipment Data Sheet, Engineering Drawing, Loading Data
  - Construction 관련

PFD Example

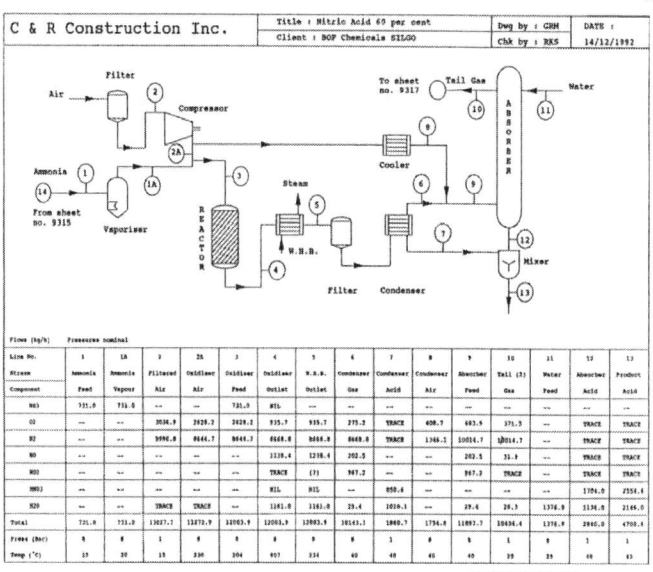

P&ID Example

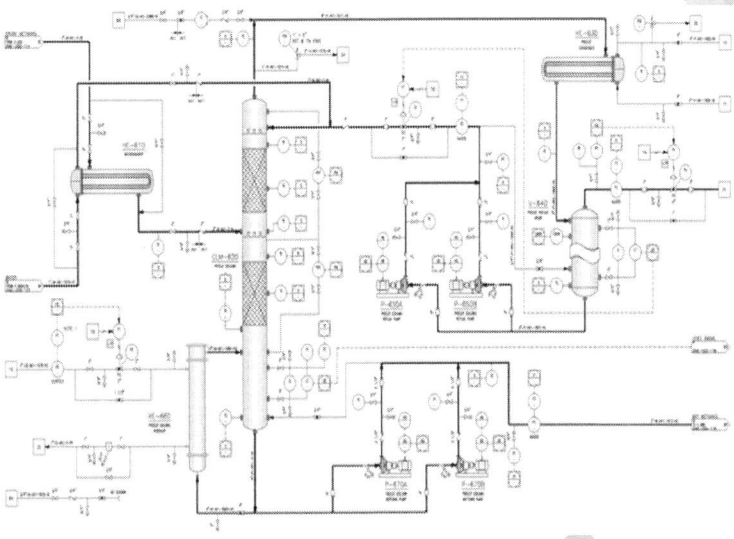

공장 용지/부지 선정

> 입지조건

(1) 원료 (2) 시장성 (3) 에너지공급 (4) 기후 (5) 수송 (6) 공업용수
(7) 노동력공급 (8) 폐기물처리 (9) 세금과 법적제한 (10) 확장가능성 (11) 생활환경

추가 고려사항 : - 과세 및 법적인 제약 여부
  - 지역사회 특성 및 영향

> 지질적 조건
  - 토질 및 지반구조 조사

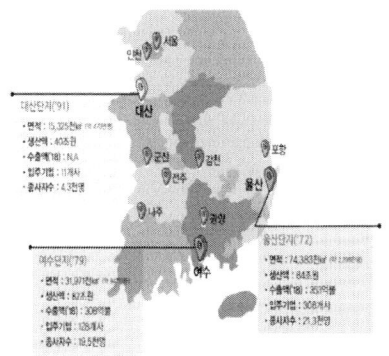

Source : 한국석유화학협회 미니북

Source : 현대/기아차 자료

➢ 일본 Oil refinery

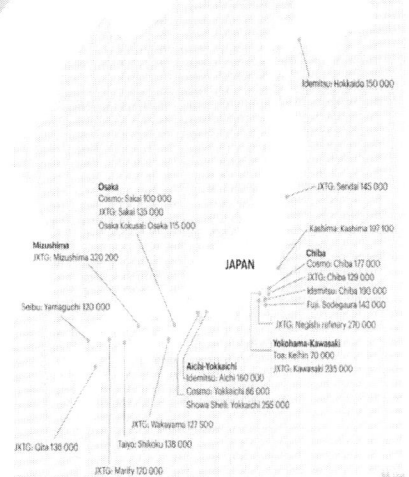

Source : www.iea.org

➢ 미국 Oil refinery

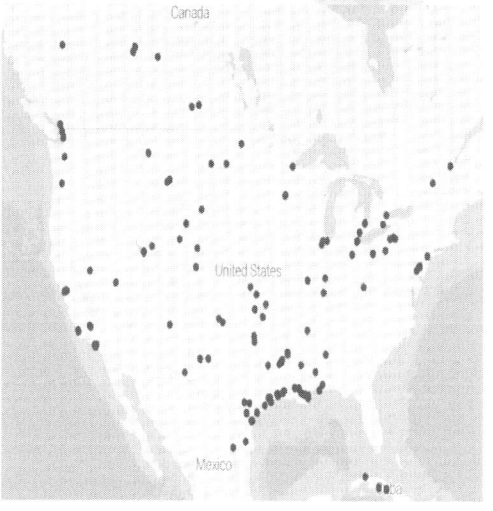

Source : www.insights-global.com

공장 배치도 (Layout)

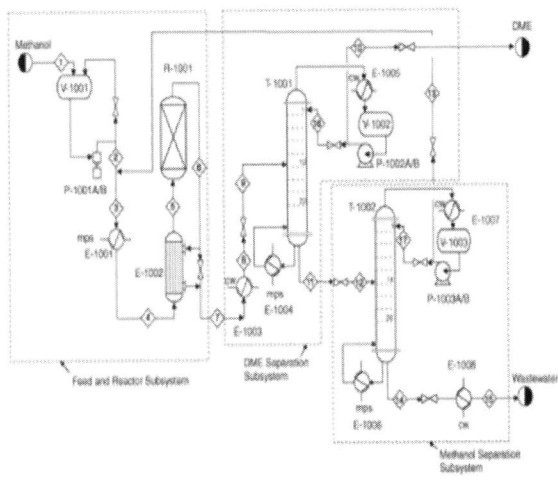

공장 배치도 (Layout)

1. Site Layout
    Relevant to neighborhood (sewage, nearby roads, populated areas etc.)

2. Plot Layout
    Actual layout of the plot, fences, general area allocation.

3. Equipment Layout

공장 배치도 (Layout)

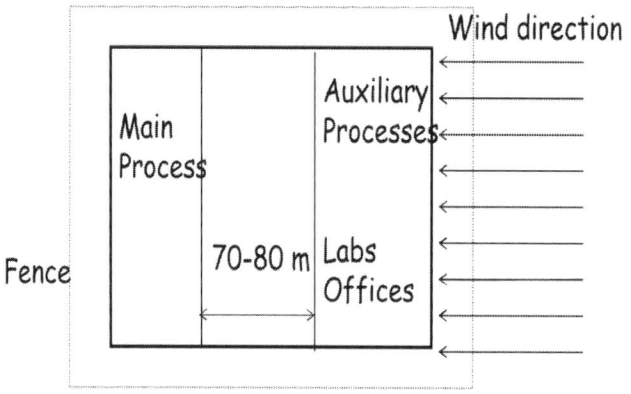

- 법정 이격 거리 준수
- Auxiliary Processes
  : Power generation, Cooling water, Cooling tower, Boiler 등
- Wind Direction ???

공장 배치도 (Layout)

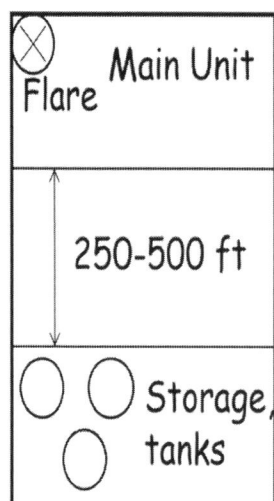

- Storage tanks
  : 저장물이 Gas/Liquid/Solid 및 압력/온도에 따른 이격 거리 준수

- Flare
  : Vent되는 Gas 성분의 폭발을 방지하기 위한 안전 설비

공장 배치도 (Layout)

Main Process :

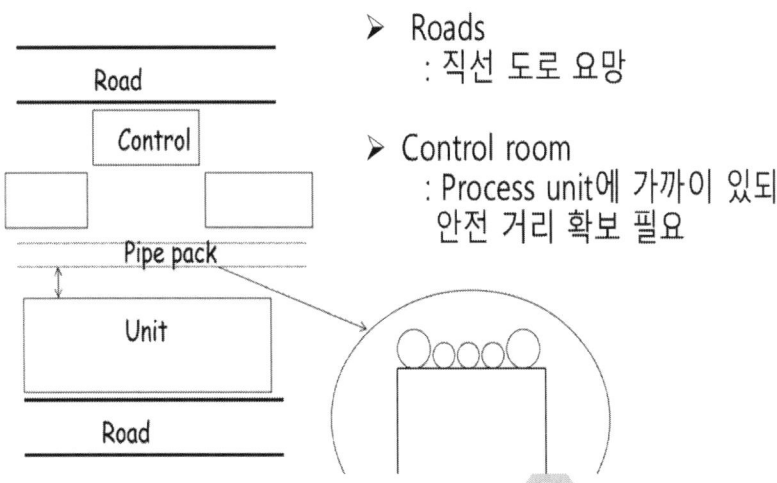

- Roads
  : 직선 도로 요망

- Control room
  : Process unit에 가까이 있되, 안전 거리 확보 필요

공장 배치도 (Layout)

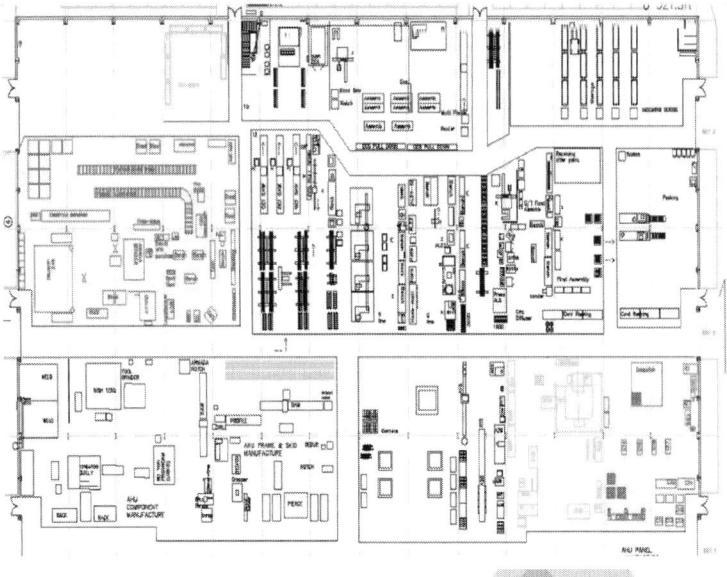

참고) Plot Layout Model

강 의 노 트

🚗 1.

🚗 2.

🚗 3.

제7장

Flow Diagram

Flow Diagram

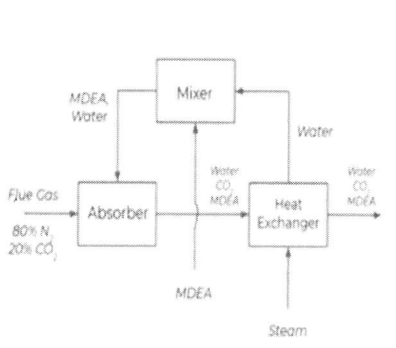

Flow Sheet 작성의 중요성

> Process design의 시작점

The Flow-sheet Importance

- Shows the arrangement of the equipment selected to carry out the process.
- Shows the streams concentrations, flow rates & compositions.
- Shows the operating conditions.

- During plant start up and subsequent operation, the flow sheet from a basis for comparison of operating performance with design. It's also used by operating personnel for the preparation of operating manual and operator training.

Block Flow Diagram (1/3)

> Sodium docecylbenzene sulfonate production

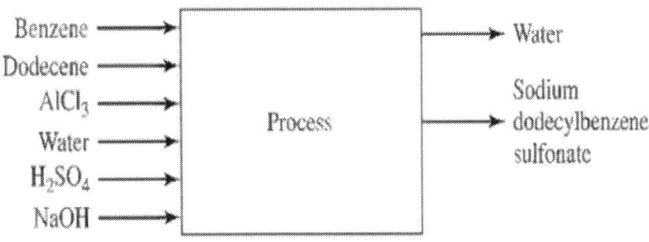

Input-Output Diagram

Block Flow Diagram (2/3)

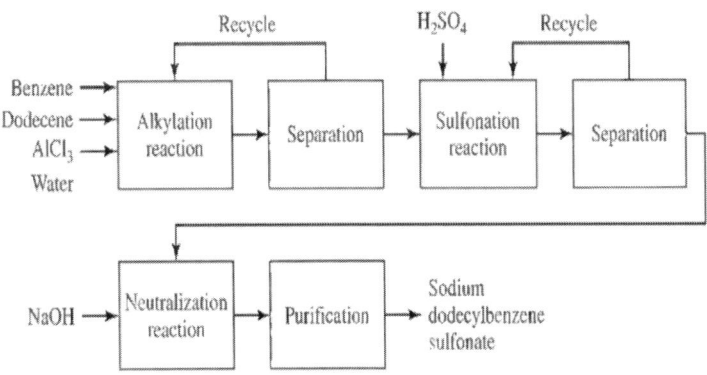

Functions Diagram

Block Flow Diagram (3/3)

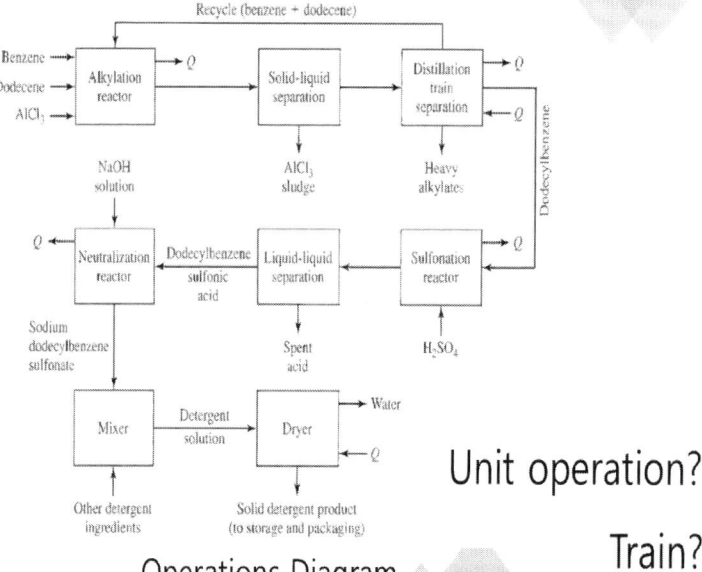

Unit operation?

Train?

Operations Diagram

Conceptual Process Diagram

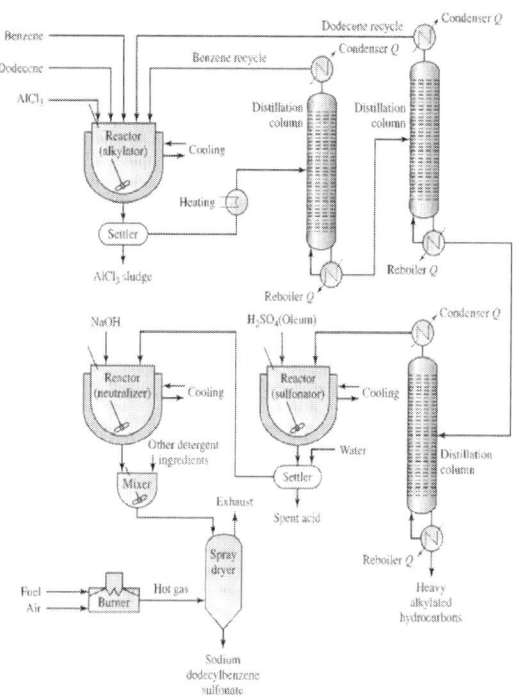

Block Flow Diagram의 활용

> BFD 정보

- Represent the process in a simplified form.
- No details involved.
- Don't describe how a given step will be achieved.

When is it used?

- In survey studies.
- Process proposal for packaged steps.
- Talk out a processing idea.

참고) 문서결재 BFD 적용 사례

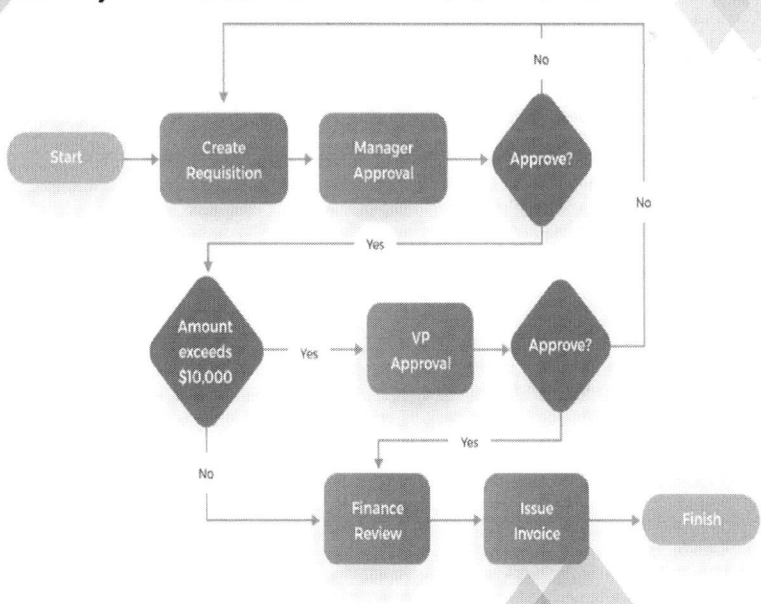

대표적인 Pictorial Flow Sheet

> 주요 장치들의 그림 또는 물질의 기본 정보들을 포함

- ❖ Process Flow Diagram (PFD)
- ❖ Piping & Instrumentation Diagram (P&ID)
- ❖ Utility Flow Diagram (UFD)

Process set-up procedure 사례 (1/10)

> Ethylene을 생산하기 위한 공정 Set-up

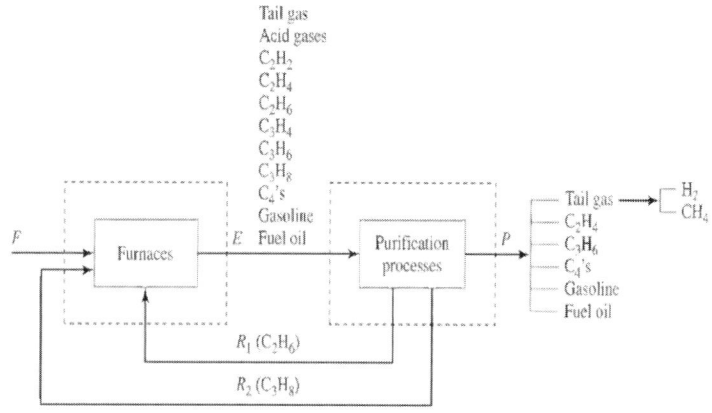

Simplified Flow Diagram

Process set-up procedure 사례 (2/10)

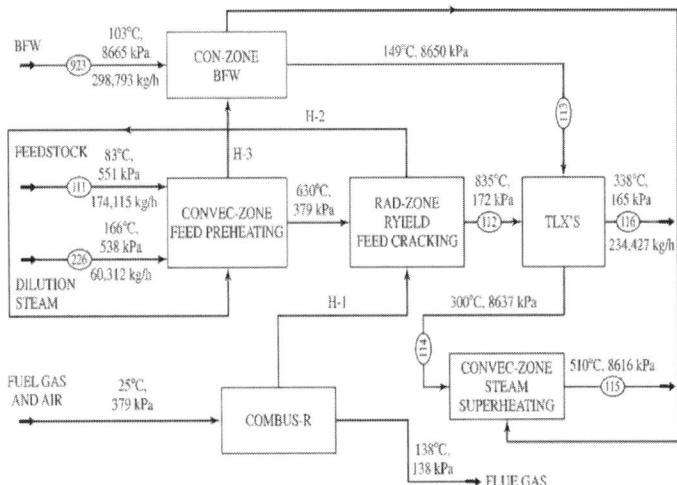

열분해로 Flow Diagram

Process set-up procedure 사례 (3/10)

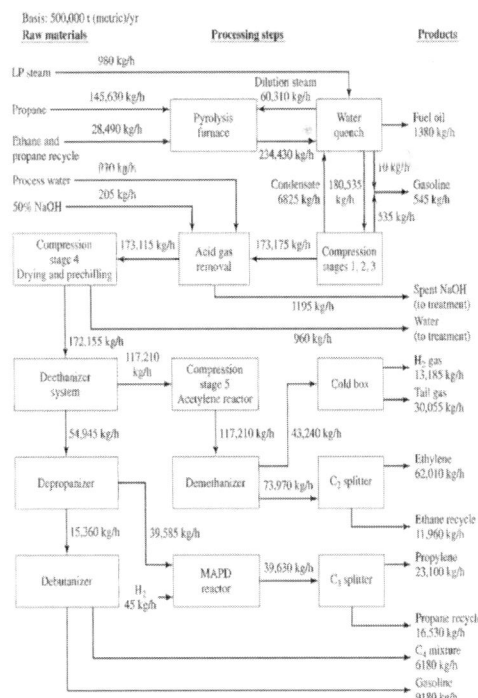

Material Balance Summary

Process set-up procedure 사례 (4/10)

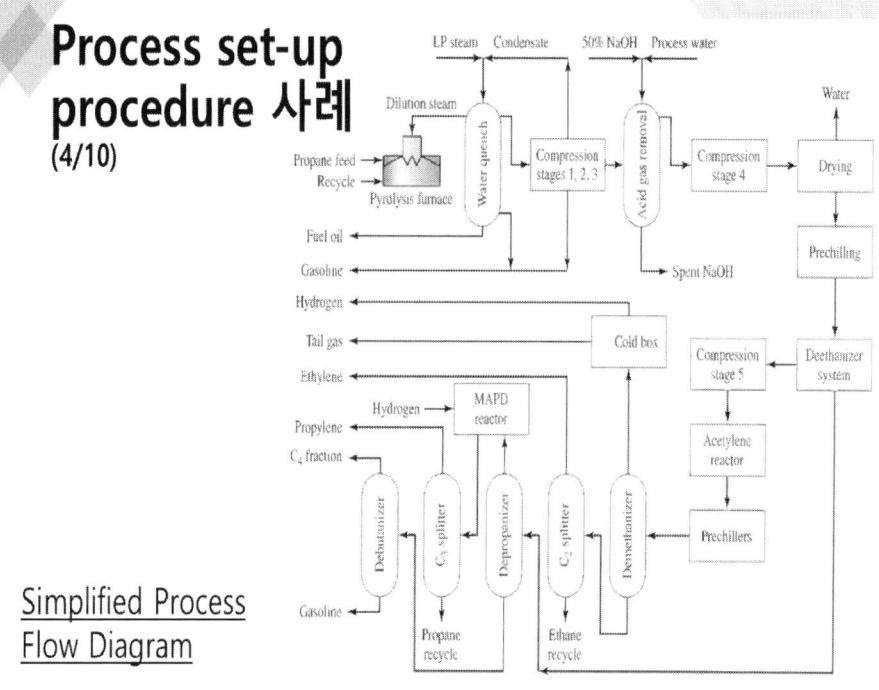

Simplified Process Flow Diagram

Process set-up procedure 사례 (5/10)

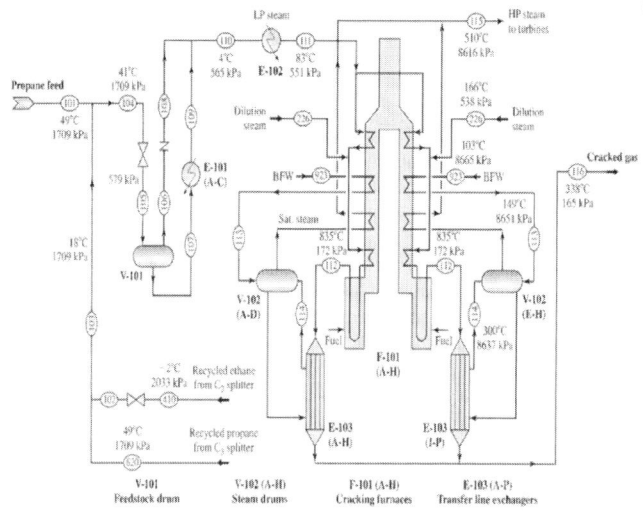

Process Flow Diagram for Cracking Section

Process set-up procedure 사례 (6/10)

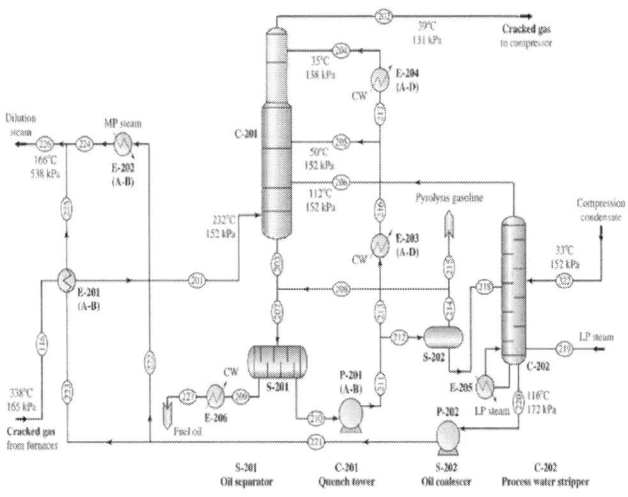

Process Flow Diagram for Quenching Section

Process set-up procedure 사례 (7/10)

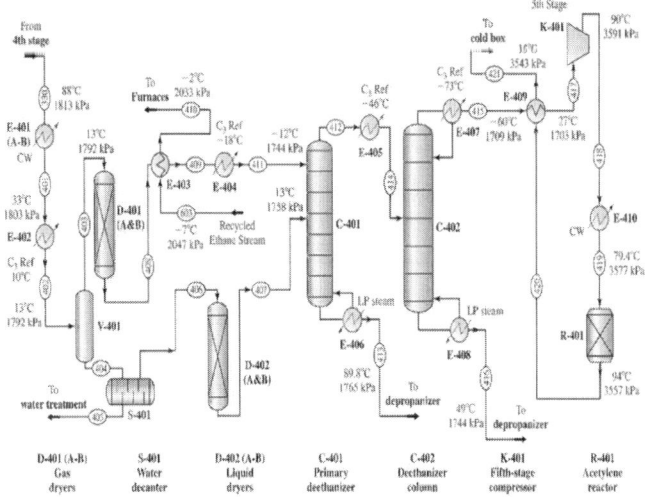

Process Flow Diagram for Acetylene Hydrogenation Section

Process set-up procedure 사례 (8/10)

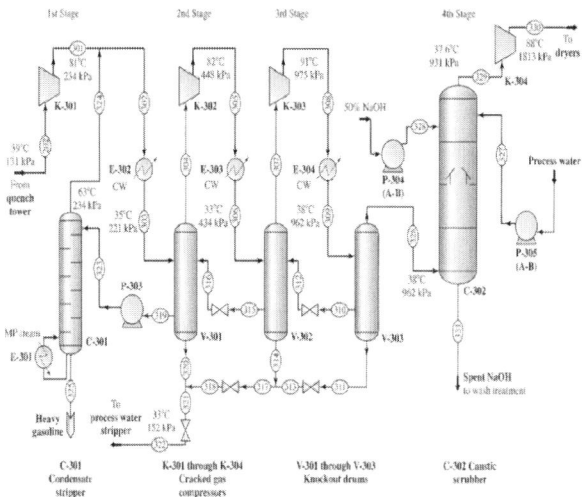

Process Flow Diagram for Compression & Acid Gas Removal Section

Process set-up procedure 사례 (9/10)

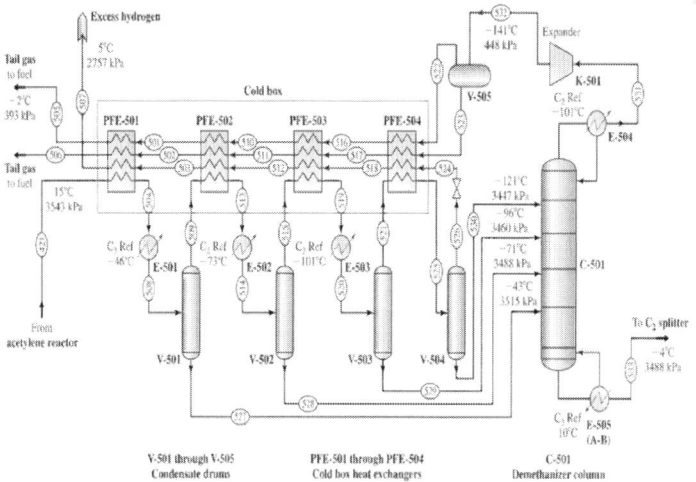

Process Flow Diagram for Chilling & Demethanization Section

Process set-up procedure 사례 (10/10)

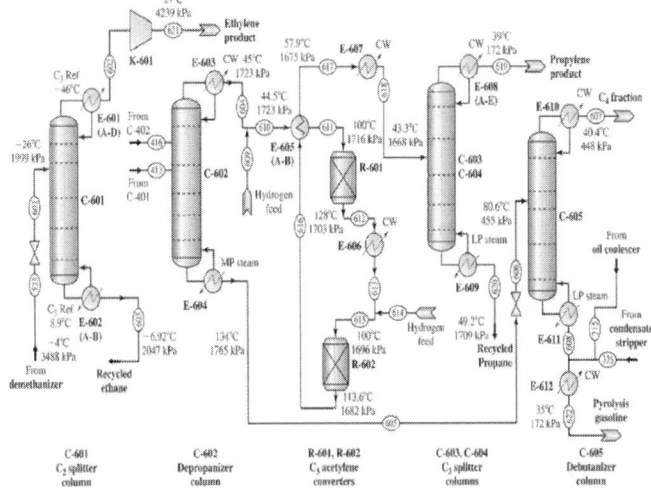

Process Flow Diagram for Product Separation Section

PFD

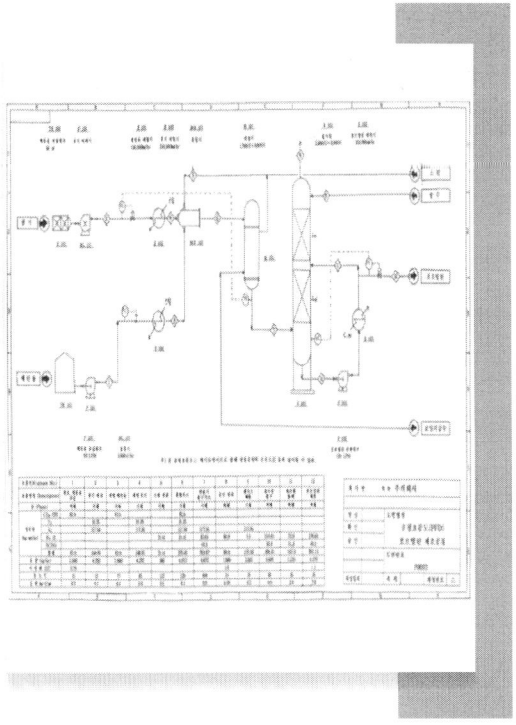

PFD

- ✓ PFD is a simplified flow diagram of a single process unit, a utility unit, and a complete process module.

- ✓ The purpose of a PFD is to provide a preliminary understanding of the process system indicating only the main items of equipment, main pipelines and the essential instruments, and valves

- ✓ PFD also indicates operating variables (mass flow, temperature and pressure), which are tabulated at various points in the process system.

PFD

- ✓ The PFD is a document containing information on :

  - Process conditions and physical data of the main process streams
  - Main process equipment with design data
  - Main process lines
  - Mass (material) balance
  - Heat balance (if applicable)

PFD Example (1/3)

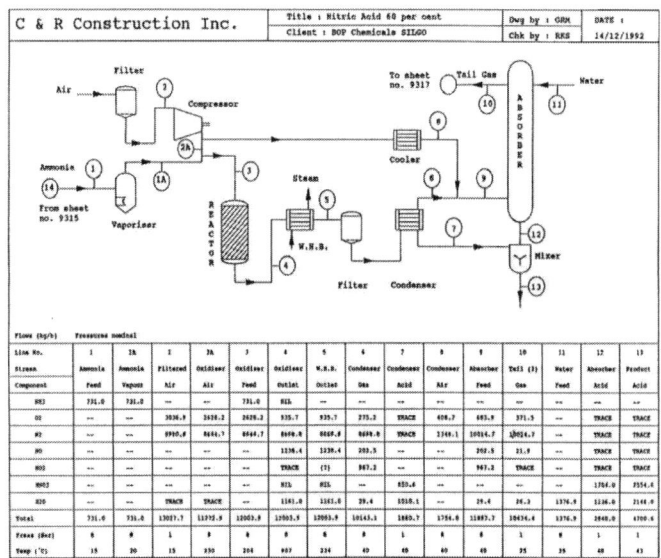

PFD Example (2/3)

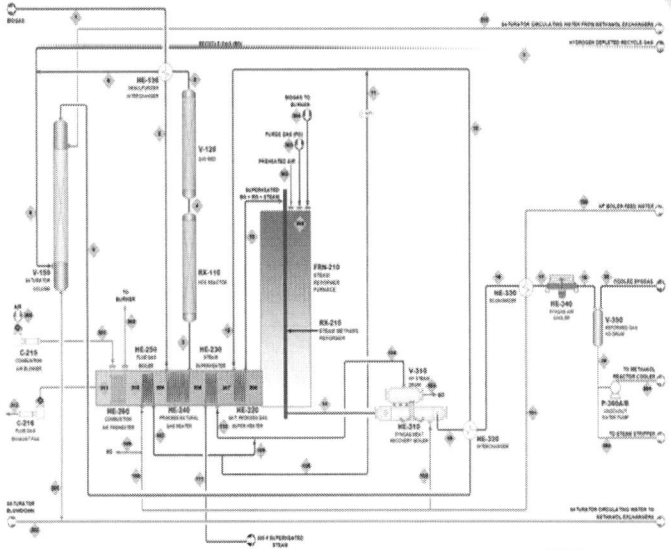

PFD Example (3/3)

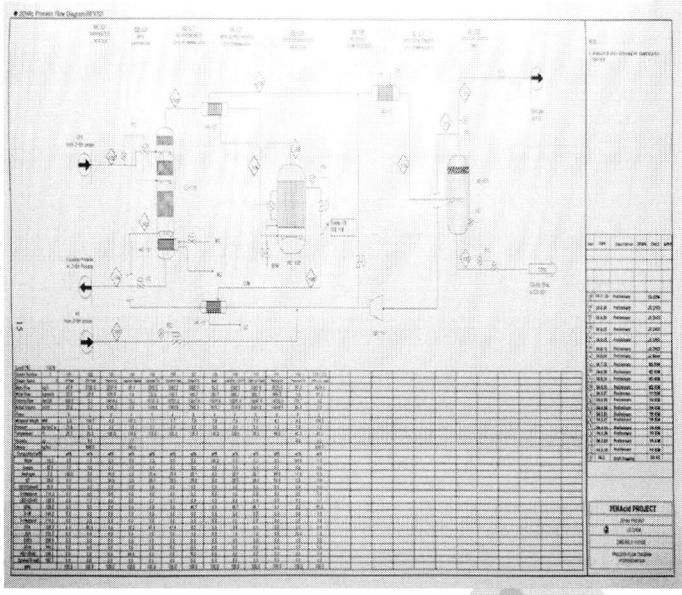

PFD details (1/3)

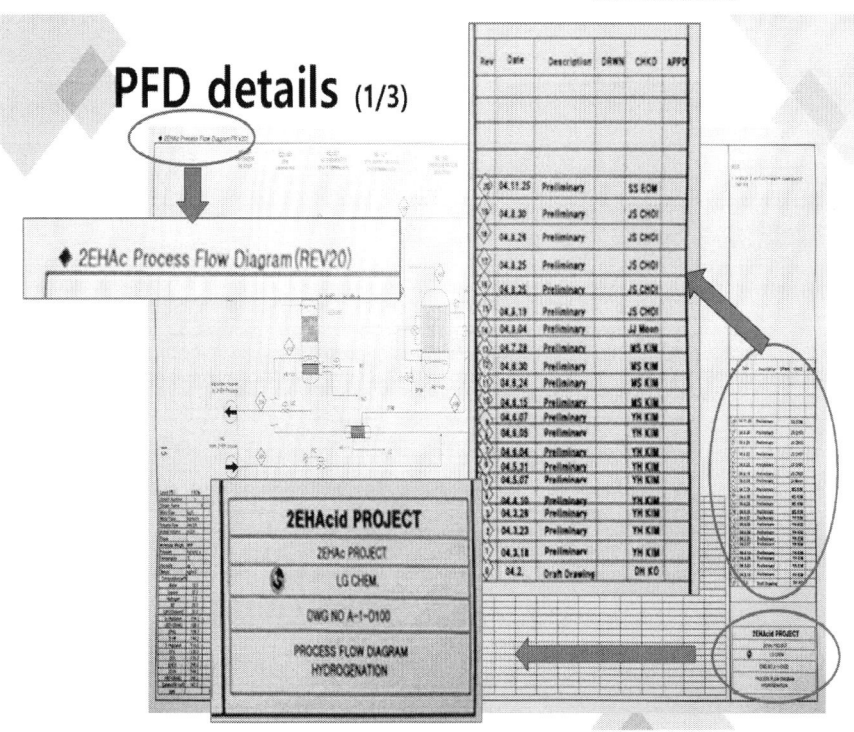

PFD details (2/3)

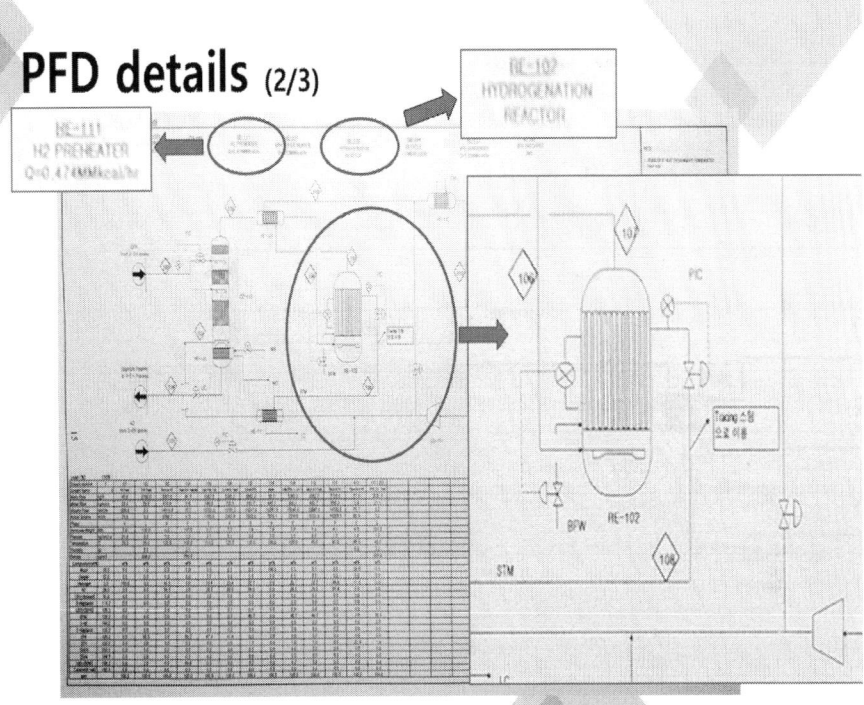

PFD details (3/3)

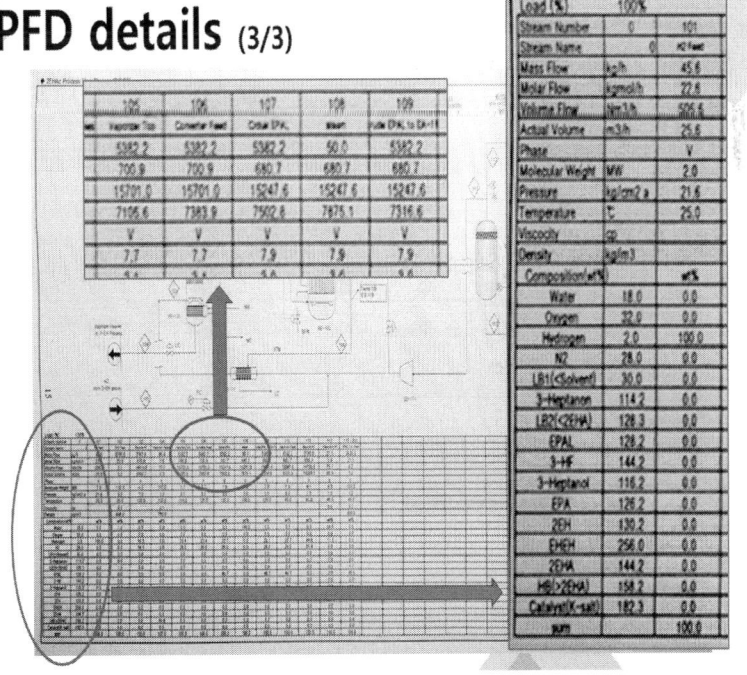

PFD 작성을 위해...

- Unit 공정의 in-out mass balance 중요, 특히 반응공정의 mass balance 확립이 key-point

- 공정 조건의 해당 물질의 physical data 확보

- 운전 조건의 설정 범위 결정 : Heat/Energy balance & integration

- Unit 공정의 연결

- Unknown material 처리

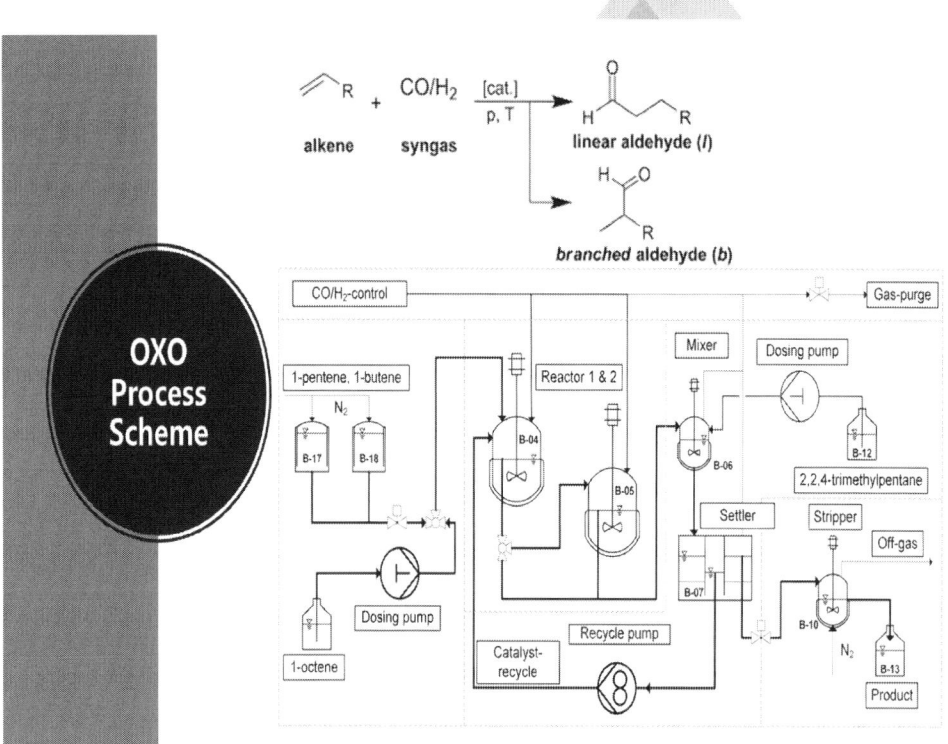

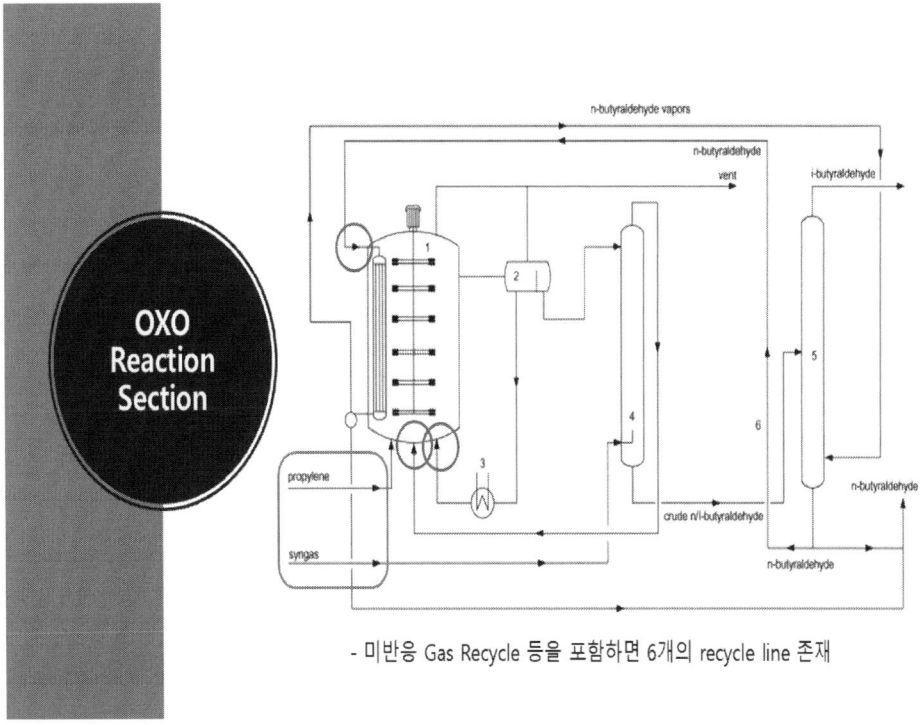

- 미반응 Gas Recycle 등을 포함하면 6개의 recycle line 존재

UFD

✓ Used to summarize and detail the interrelationship of utilities, such as air, water (various types), steam (various types), heat transfer mediums, process vent and purges etc.

✓ UFD is also containing information on :

- Main distribution or arrangement of each individual utility system, expect electrical system

> Consideration for utilities

❖ Water : Process water, cooling water, production of steam, fire water, drinking water...

❖ Air : Process air, instrument application air, fuel air, purge air...

참고) Process water?

✓ Process Water는 물 속의 불순물을 제거하여 해당 공정에 적합한 물로 공정 운전에 영향을 미치지 않도록 하기 위함

❖ Boiler Feed Water : 처리된 물이 아닌 경우,
　　　　　　　　　　Boiler 표면이나 응축수 line의 부식 또는
　　　　　　　　　　침전물 생성으로 Boiler 효율 저하

> Boiler Feed Water(BFW)
　: Demineralized water에서 산소를 제거한 water
　(* Demineralized water : 염분을 제거한 water (Demi-water로 호칭))

- Brine Water : Chloride ion이 1,000~10,000mg/L 포함된 소금물,
  - 냉각기 등의 냉매로 사용

참고) Boiler Feed Water Spec. Example

Constituent	Units	Feedwater	Product (HPBFW)
Calcium (Ca)	mg/l	45	
Magnesium (Mg)	mg/l	11	
Sodium (Na)	mg/l	61	<0.010
Chloride (Cl)	mg/l	74	
Sulfate (SO_4)	mg/l	60	
Bicarbonate (HCO_3)	mg/l	150	
Silica	mg/l as SiO_2	45 (base), 0-150	<0.010
pH	-	7.0	
TDS (non-silica)	mg/l	401	
TDS (total)	mg/l	446	
TDS	mg/l as $CaCO_3$	290	
Conductivity	µS/cm		<0.1

Source : www.globalwaterintel.com/sponsored-content/performance-under-pressure-high-quality-boiler-feedwater

참고) Instrument Air?

✓ 주로 Plant 에서 공기식 계기와 조절기 등의 작동용 및 제어 신호의 전달매체로 사용되는 Air

✓ Compressor에서 압축한 후, Dryer를 통과하여 얻은 건조된 Air

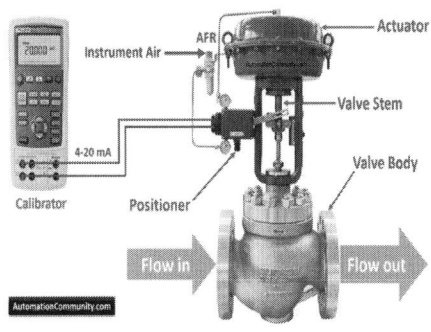

UFD Example

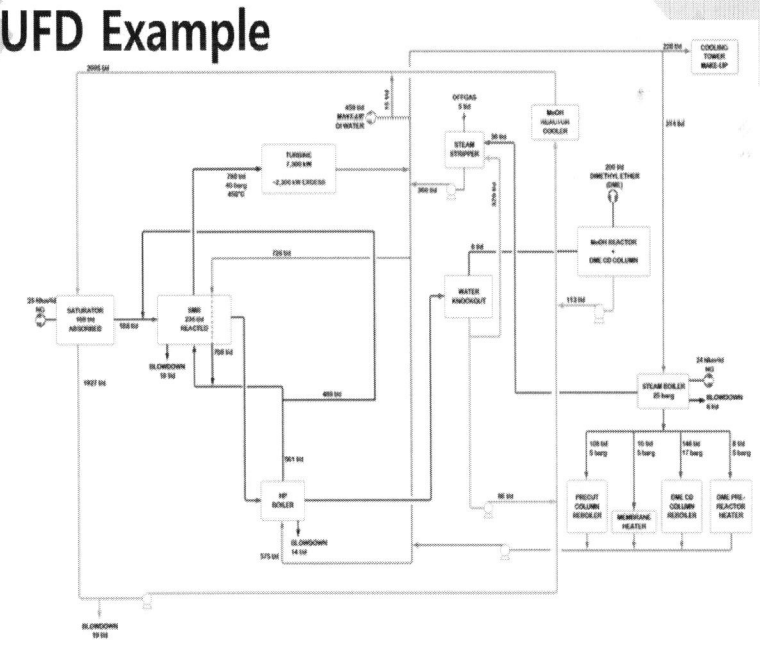

P&ID

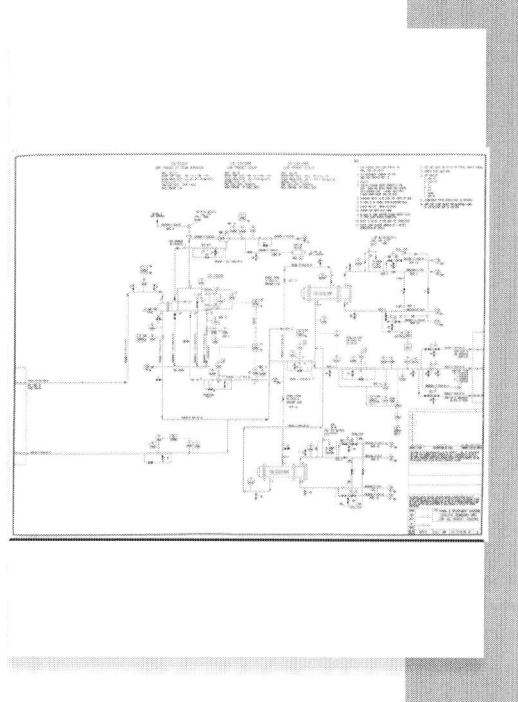

P&ID

✓ P&ID shows the arrangement of the process equipment, piping, pumps, instrument, valves and other fittings.

✓ P&ID should include :

- All process equipment identified by an equipment number
- All pipes identified by a line size, material code and line number
- All valves with an identified size and number
- All pumps identified by a suitable code number
- All control loops and instruments

P&ID Example (1/2)

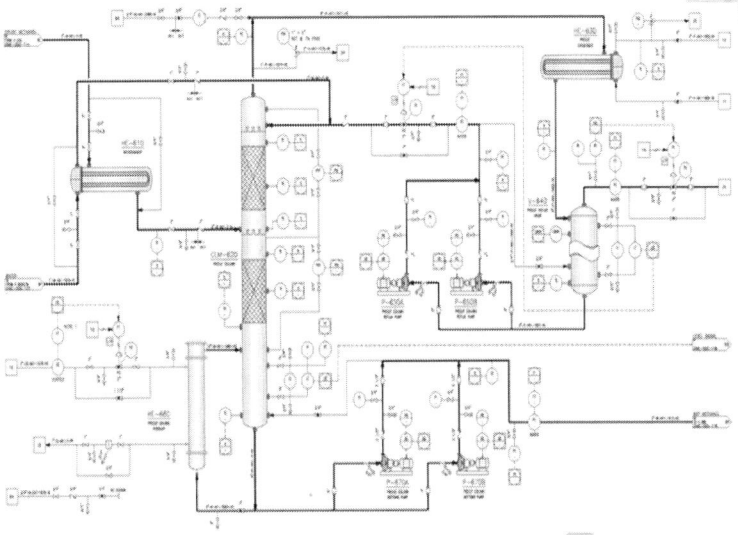

P&ID Example (2/2)

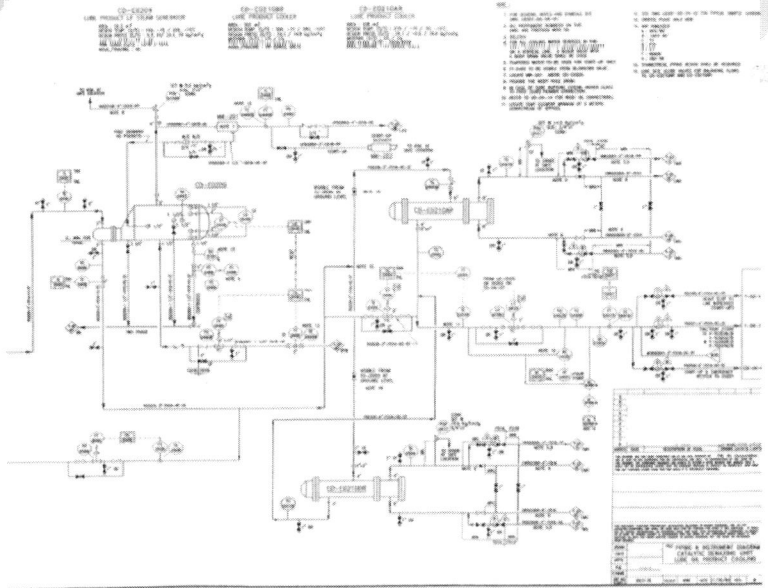

P&ID 표기 방법 및 의미 (1/2)

✓ 장치 및 동력기계 고유번호 부여방법

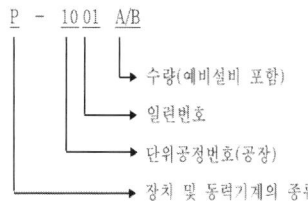

P : 펌프류	C : 압축기류	
AG : 교반기류	HT : 호이스트류	
T : 탑류	R : 반응기	
D : 드럼류	TK : 저장탱크류	
E : 열교환기류	H : 히타류	
V : 용기류	X : 공급자 일괄공급기기	

P&ID 표기 방법 및 의미 (2/2)

✓ 배관번호 부여방법

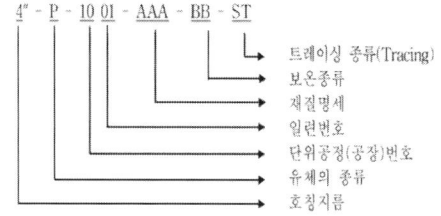

✓ 계기의 고유번호 부여방법

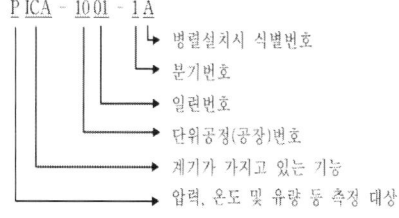

P&ID Symbol (1/4)

✓ Pumps

P&ID Symbol (2/4)

✓ Valves

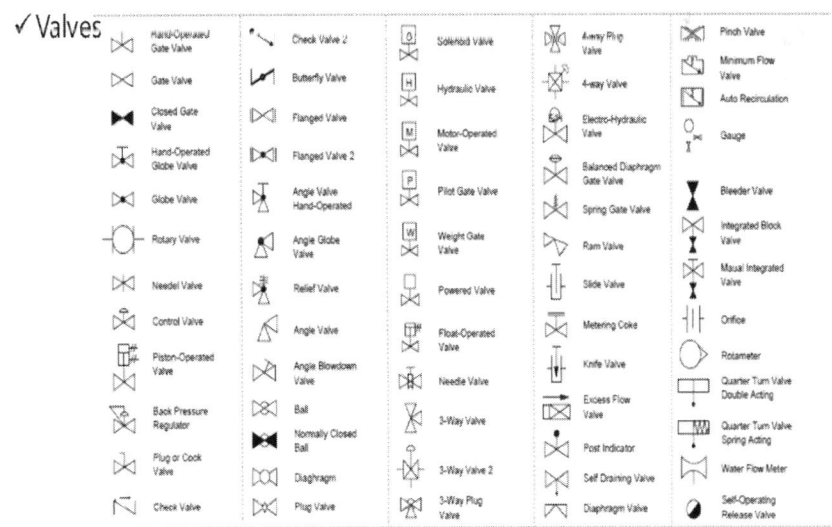

P&ID Symbol (3/4)

✓ Instruments

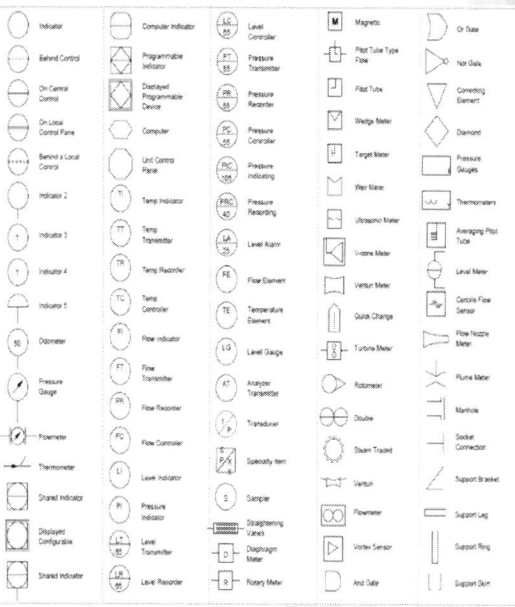

P&ID Symbol (4/4)

✓ Piping & Connecting

Equipment List & 기타

	Rock Drill	3	Drill rock to facilitate excavation of the trench
	Sandblaster	5	Clean the pipe welds prior to coating
ROW Preparation, Pipe Installation, Backfill, and Cleanup (cont.)	Sideboom	15	Lower pipe
	Street Sweeper	4	Clean mud and debris from roadways
	Stringing Truck	10	Transport pipe to the ROW
	Water Truck	20	Dust control
	Welding Truck	80	Weld pipe
	Wheel Loader	3	Load materials into dump trucks
Paving	Dump Truck	16	Transport asphalt
	Grinding Machine	4	Prepare asphalt for paving
	Paving Machine	4	Resurface roads following installation of the pipe
	Pick-up Truck	4	Transport workers and materials
	Steamroller	4	Resurface roads following installation of the pipe
	Street Sweeper	4	Clean mud and debris from roadways
Civil Work	Air Compressor	5	Power pneumatic tools
	Flatbed Truck	5	Transport construction materials
	Pick-up Truck	5	Transport workers and materials
Horizontal Bore	Air Compressor	4	Power pneumatic tools
	Boring Machine	4	Drill beneath roadways, sensitive resources, and/or existing utilities
	Crane	4	Lower the boring machine into the bore pit
	Flatbed Truck	4	Transport materials to the bore site
	Pick-up Truck	4	Transport workers and material

Equipment List

✓ Equipment List shows the specifications of each equipment in the process

✓ Equipment List is a document containing information on :

- Equipment Item number, service name
- Material type
- Size, Volume, Height
- Design Temperature and Pressure

Equipment List Example (1/3)

EQUIPMENT LIST

REACTOR & VESSEL

ITEM NO.	SERVICE	QTY	TYPE	MATERIAL	Vol. (m3)	Press. (kg/cm² G)	Temp. (℃)	REMARKS
RE-102	Hydrogenation reactor	1	Tubular	CS		9.0 ~ F/V	200~10.8	30mm id*6000mm L*pich105*1,200ea
RE-301	Oxidation reactor	1	V	SUS304L	15	Rating=#600		산화반응으로 폭발위험
RE-503	Catalyst mix vessel	1	V	Sus304	3.8	1.0 ~ F/V	200	
TK-504	2EHAc slop tank	1	CRT	Sus304	100	560 ~ -60mmAq	160	

COLUMN

ITEM NO.	SERVICE	QTY	TYPE	No. of tray	Dia. 1 (mm)	Dia. 2 (mm)	Height 1 (mm)	Height 2 (mm)	Total H (mm)	MAT	Press. (kg/cm² G)	Temp. (℃)
CO-101	EPA vaporizer	1								CS	8	-10.8~200
CO-201	EPAL column	1	Packed	10	1690	1750	5486.4	609.6	6096	Sus304	5.0 / FV	225
CO-401	Fore column	1	Packed	10	1190		6096		6096	Sus316L	5.0 / FV	225
CO-402	Refining column	1	Packed	10	470	510	2438.4	3657.6	6096	Sus316L	5.0 / FV	225
CO-403	Product column	1	Packed	10	1710		6096		6096	Sus316L	5.0 / FV	225

Equipment List Example (2/3)

COMPRESSOR

ITEM NO.	SERVICE	QTY	TYPE	Material	Flow		S. Press.	Dis. Press.
			ANSI	Casing	kg/h	m3/h	(kg/cm².a)	(kg/cm².a)
GB-104	Recycle compressor	1	Centrifugal	CS	4695	5962	4.2	5.2
GB-501	EPAL column vacuum pump	1	Liq.Ring	Sus316L	200		0.1	1.15
GB-502	Fore/Refining/Product column vacuum pump	1	Liq.Ring	Sus316L	500	386 nm3/h	0.1	1.15
GB-506	Air blower	1	Turbo Fan	Sus304		200	-60mmaq	440mmaq

HEAT EXCHANGER

ITEM NO.	SERVICE	QTY	TYPE	Material		H.D	HTA	Press.	Temp.
				Shell	Tube	(mmk/h)	(m2)	(kg/cm² G)	(℃)
HE-111	H2 preheater	1	S&T	CS	CS	0.474	17.28	2.5	24
HE-112	VPH condenser	1	S&T	CS	CS	1.05	28.33	5 / 2.1	-10 / 145
HE-121	VPH super heater	1	S&T	CS	CS	0.030	2.74	2.1	150
HE-131	vaporizer heater	1	S&T	CS	CS	0.616	19.4	2.2	150
HE-211	EPAL column side cooler	1	S&T	CS	Sus304	0.050	4.17	4 / -0.81	32 / 112
HE-221	EPAL column condenser	1	S&T	CS	Sus304	1.015	90.12	4 / -0.816	32 / 81
HE-231	EPAL column reboiler	1	S&T	CS	Sus304	1.148	238.7	15 / -0.802	200 / 132.5
HE-311	Oxidation reactor cooler	1	S&T	CS	Sus304	1.320	118.4	4 / 1.97	32 / 58.6
HE-421	Fore column condenser	1	S&T	CS	Sus316L	0.160	27.6	4 / 0.999	30 / 124

Equipment List Example (3/3)

PUMP									
ITEM NO.	SERVICE	QTY	TYPE	Material	Flow	Head	Dis. Press	Temp.	Power
			ANSI	Casing	m3/h	(m)	(㎏/㎠ G)	(℃)	(kw)
PU-211	Oxidation feed pump	2	Canned	Sus304	5.03	34.92	2	50	1.22
PU-221	EPAL column reflux pump	2	Magnetic	Sus304	31	80	6.5	100	M
PU-231	EPAL column btm pump	2	Magnetic	Sus304	4	80	6.0	130	M
PU-311	Oxidation reactor circulation pump	2	Canned	Sus304	409	0.5	2.0	60	0.53
PU-411	Fore column side pump	2	Magnetic	Sus316L	4	50	5.0	100	M
PU-421	Fore column Reflux pump	2	Canned	Sus316L	4	50	5.0	50	M
PU-422	Refining column reflux pump	2	Canned	Sus316L	4	50	5.0	50	M
PU-423	Product column reflux pump	2	Magnetic	Sus304	4	50	5	130	M
PU-431	Fore column btm pump	2	Magnetic	Sus316L	4	50	7.0	180	M
PU-432	Refining column btm pump	2	Magnetic	Sus316L	4.0	50	5.0	150	M

✓ 총 투자비 산출 가능

Operation manual & 기타 사항

✓ SOP : Standard Operation Procedure

✓ Start-up Procedure & Shut-down Procedure

✓ Emergency Shut-down Procedure

✓ Sampling Point & Sampling Method

✓ Analysis Method & Specification

✓ Chemical & Physical Data

✓ Etc.

Equipment Data Sheet Example

DISTILLATION COLUMN DATA SHEET		Tag No	T-100
		Sheet No	1 of 1
		Function	Removal of cyclohexane
Operating Data			
COLUMN INSIDE DIAMETER (m)	2.080	STRIPPING TRAYS	10 TRAYS
COLUMN OUTSIDE DIAMETER (m)	2.086	RECTIFYING TRAYS	6 TRAYS
COLUMN HEIGHT (m)	11	NO OF TRAYS	16 TRAYS
TRAY SPACING (m)	0.45		
Internal Condition			
	FEED	TOP	BOTTOM
TEMPERATURE (°C)	96.5	80.95	164
PRESSURE (kPa)	101.3	106.3	111.3
MOLAR FLOW (kmol/hr)	183522.820	118201.286	65321.530
MASS FLOW (kg/hr)	16394400	9949210	6445190
MOLECULAR WEIGHT (kg/kmol)	89.33167003	84.1731	98.665
DENSITY (kg/m³)	839.253	716.519	812.366
VISCOSITY (kg/ms)	0.624	0.406	0.354
VOLUMETRIC FLOW (m³/hr)	5.426	3.857	2.203

Technical / Mechanical Data			
FEED NOZZLE THICKNESS (mm)	9.8	BOLT AREA (mm)	0.0186
VAPOR NOZZLE THICKNESS (mm)	7.603	DEAD WEIGHT (kN)	425.432
LIQUID NOZZLE THICKNESS (mm)	6.319	PLATE THICKNESS (m)	0.005
COLUMN INSIDE DIAMETER (m)	2.080	GASKET INSIDE DIAMETER (m)	2.096
COLUMN OUTSIDE DIAMETER (m)	2.086	GASKET OUTSIDE DIAMETER (m)	2.100
CORROSION ALLOWANCE (mm)	2.00	SIEVE TRAY HOLE DIAMETER (m)	0.006
VESSEL MATERIALS	STAINLESS STEEL	WEIR LENGTH (m)	1.560
TRAY MATERIALS	STAINLESS STEEL	WEIR HEIGHT (m)	0.75
SKIRT MATERIAL	STAINLESS STEEL	HOLE AREA (m²)	0.000113
TRAYS FROM TOP TO BOTTOM	16 TRAYS	DOWNCOMER AREA (m²)	0.408
NET AREA (m²)	2.990	NO OF HOLES	3008
Date of enquiry			

Equipment Name Plate Example (1/2)

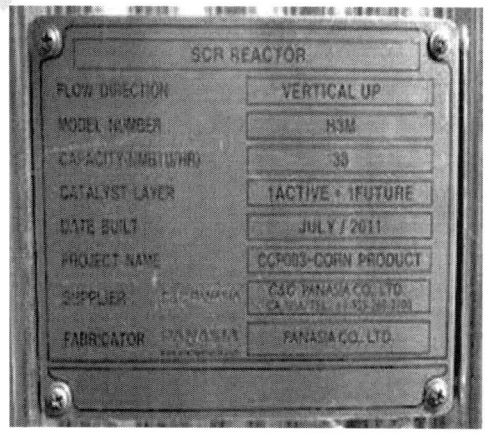
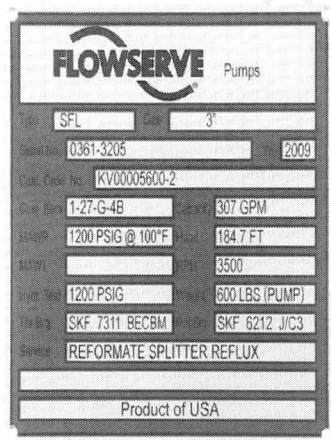

Equipment Name Plate Example (2/2)

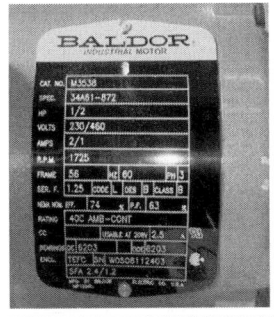

Eng. Draw. Example

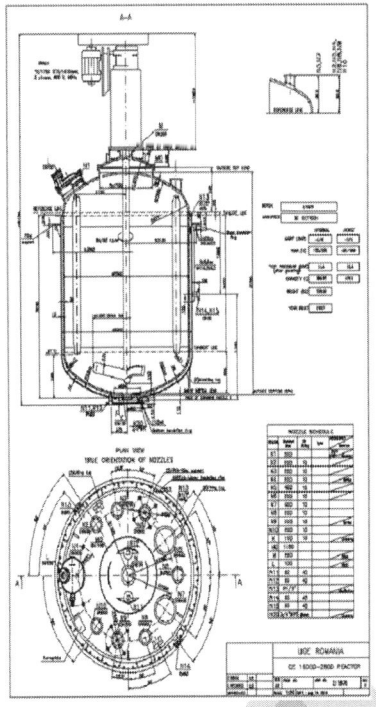

Loading Data Example

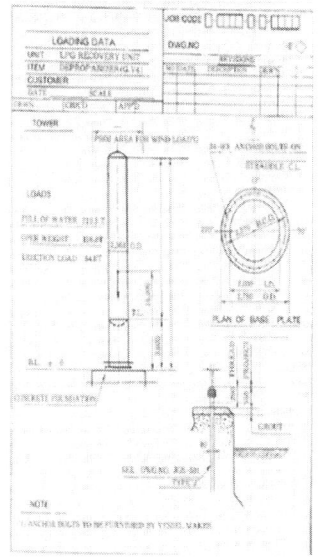 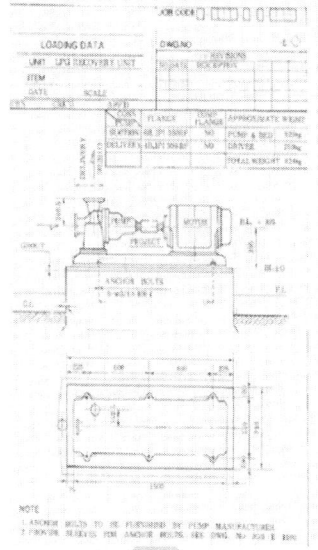

강 의 노 트

1.

2.

3.

제8장
화학 공정 반응기

Acrylic acid Process

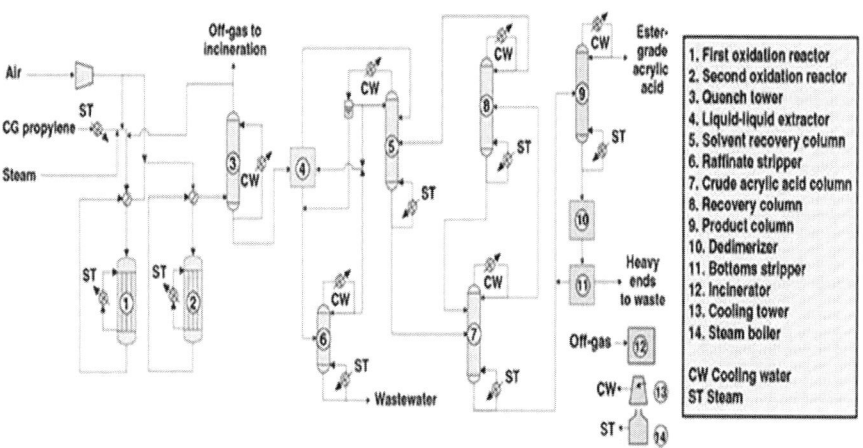

Source : Chemical Engineering Feb 2016

Acrylonitrile Process

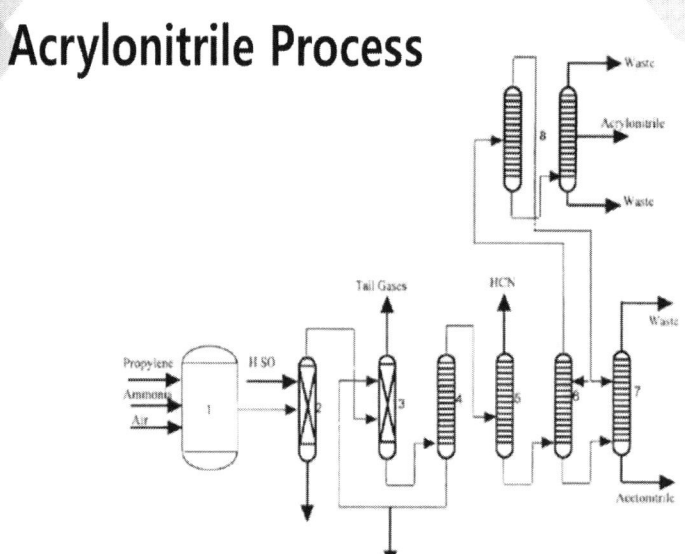

1) Reactor 2) Neutralizer 3) Absorber 4) Recovery 5) HCN Column 6) Extractive Distillation Column 7) Acetonitrile Purification Column 8) Acrylonitrile Purification Columns

Source : Guidebook of Acrylonitrile by propylene ammoxidation, 2013

Ethylene production

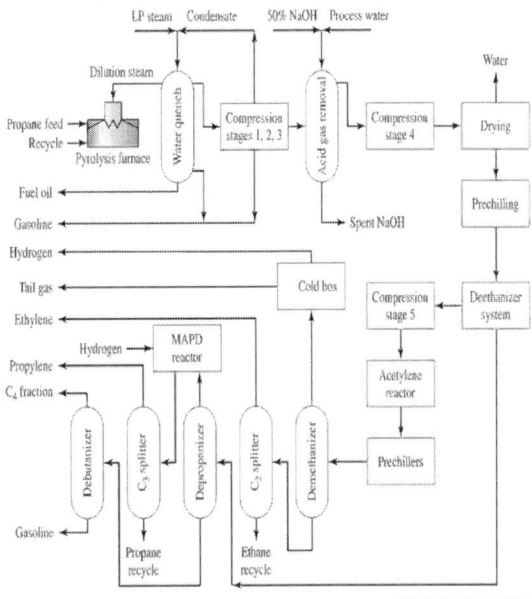

화학공학의 독특한 2개 분야

> Reaction : 반응공학   > Distillation : Transport Phenomena

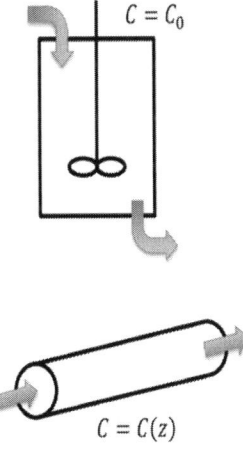

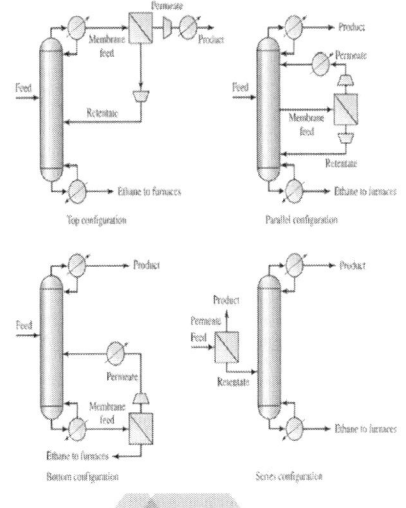

상업용 반응기

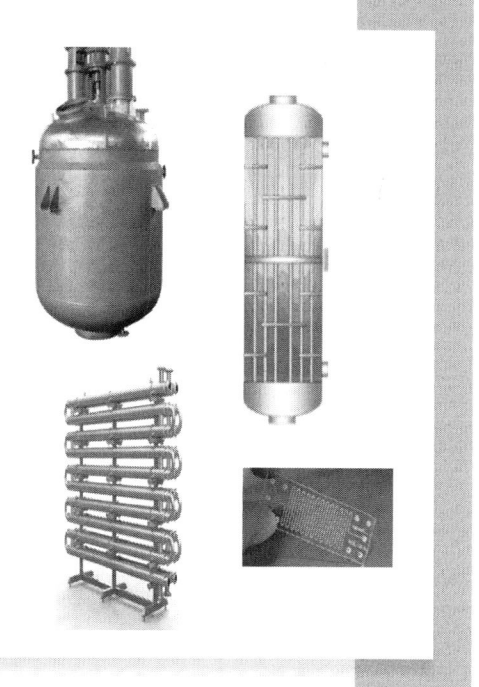

Batch Reactor

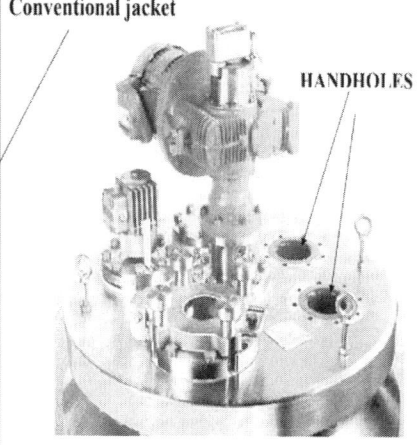

CSTR (Continuous Stirred Tank Reactor)- Cut view and impellers

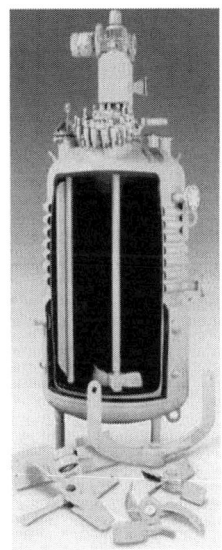

Source : www.jeiopi.co.kr/English/prd/impeller.thm

Glass-Lined Reactor

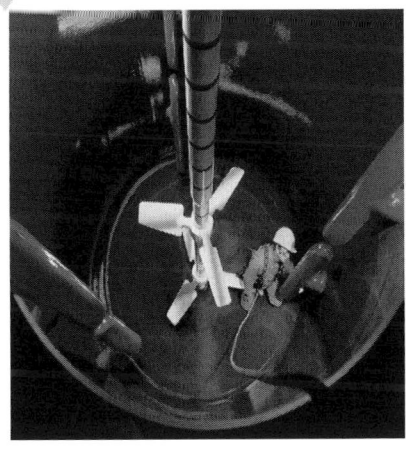
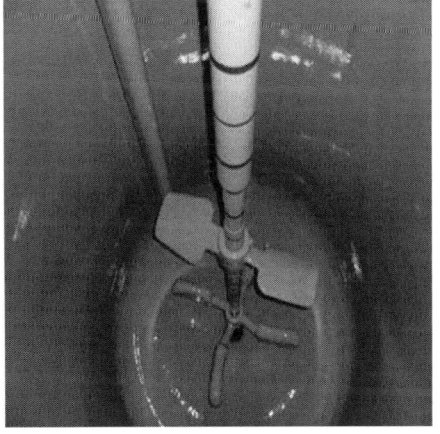

Source : www.ddpcsinc.com/glass-lined-equipment   Source : www.flexachem.cmo/mixing-technology/glass-lined-reactor

CSTR / Batch Reactor (Mixing & Impellers)

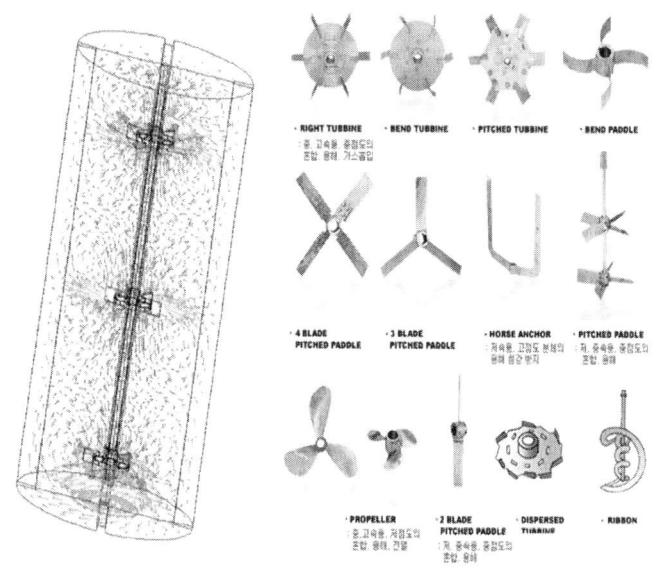

Agitator Selection Guide

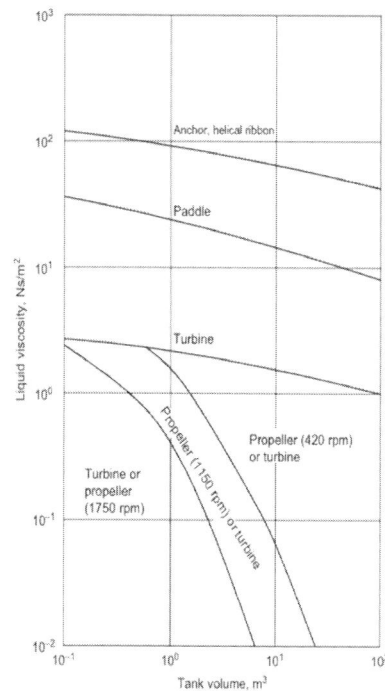

CSTR / Batch Reactor (various types of jackets)

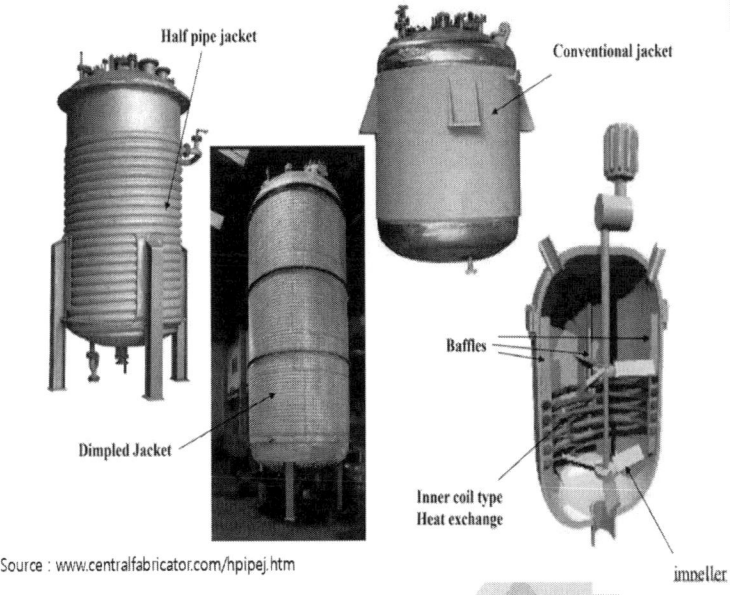

Source : www.centralfabricator.com/hpipej.htm

CSTR / Batch Reactor (on-site photo)

◆ 제8장. 화학 공정 반응기 ▶ 181

CSTR for Polymerization

Tubular Reactor for Polymerization

- ExxonMobil's tubular reactor technology for high-pressure LDPE (low density polyethylene) process

- Production capacity
  : 220,000 ton/yr (27.5 ton/hr)

Tubular Reactor for SCWO

- SCWO plant (supercritical water oxidation)

- Capacity
  : 9,000 ton/yr
    (1,100 kg/hr)

Tubular Reactor

- Microreactor

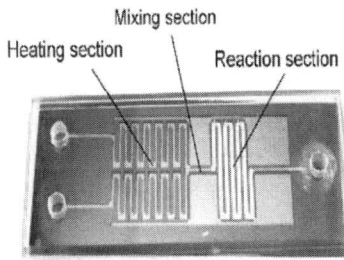

Microreactor made of silicon anodically bonded with glass

- Lab-on-Chip

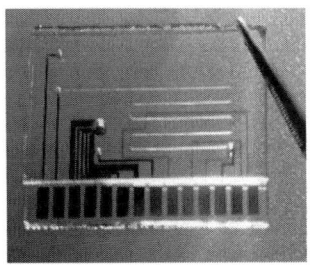

Lab-on-Chip made of glass and polymer for DNA amplification and detection

CSTR and Tubular in series

> Propylene Dimerization to isohexanes (Dimersol G unit)

Source : www.ifp.fr/dimerization-propylene

Multi-tubular Fixed Bed Reactor

> Ethylbenzene production

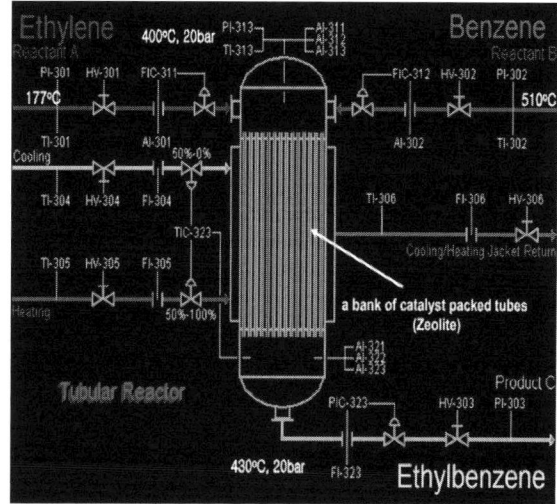

Source : www.simtronics.com/catalog/spm/spm2200a.htm

Fixed Bed Reactor (자동차 배기가스 정화장치)

$$2NO \rightarrow N_2 + O_2$$
$$2NO_2 \rightarrow N_2 + 2O_2$$
$$2CO + O_2 \rightarrow 2CO_2$$

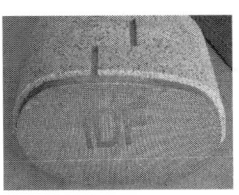
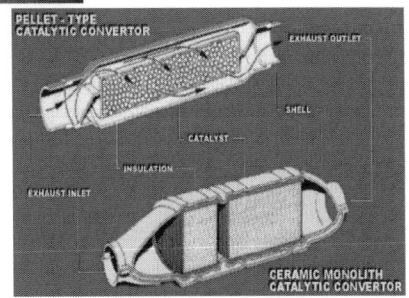

Packed Bed Reactor (Hydrotreating Unit)

Catalytic hydrotreating is a hydrogenation process used to remove about 90% of contaminants such as nitrogen, sulfur, oxygen, and metals from liquid petroleum fractions. These contaminants, if not removed from the petroleum fractions as they travel through the refinery processing units, can have detrimental effects on the equipment, the catalysts, and the quality of the finished product. Typically, hydrotreating is done prior to processes such as catalytic reforming so that the catalyst is not contaminated by untreated feedstock. Hydrotreating is also used prior to catalytic cracking to reduce sulfur and improve product yields, and to upgrade middle-distillate petroleum fractions into finished kerosene, diesel fuel, and heating fuel oils. In addition, hydrotreating converts olefins and aromatics to saturated compounds.

Trickle Bed Reactor

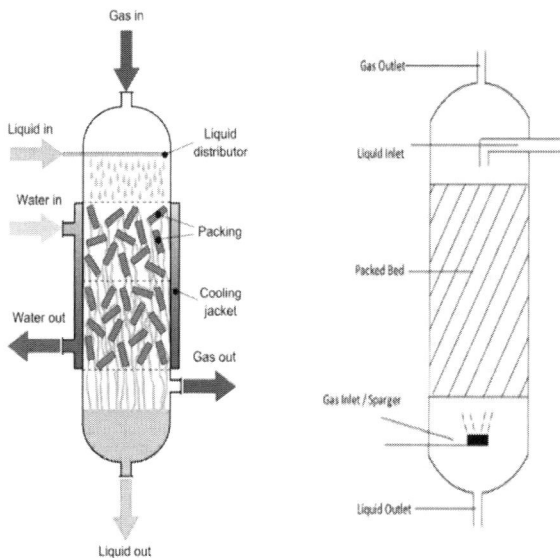

Fluidized Reactor (FCC in petroleum refining)

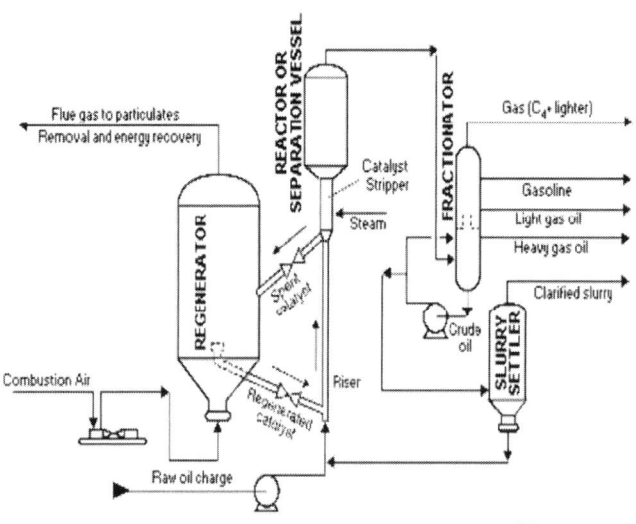

FCC

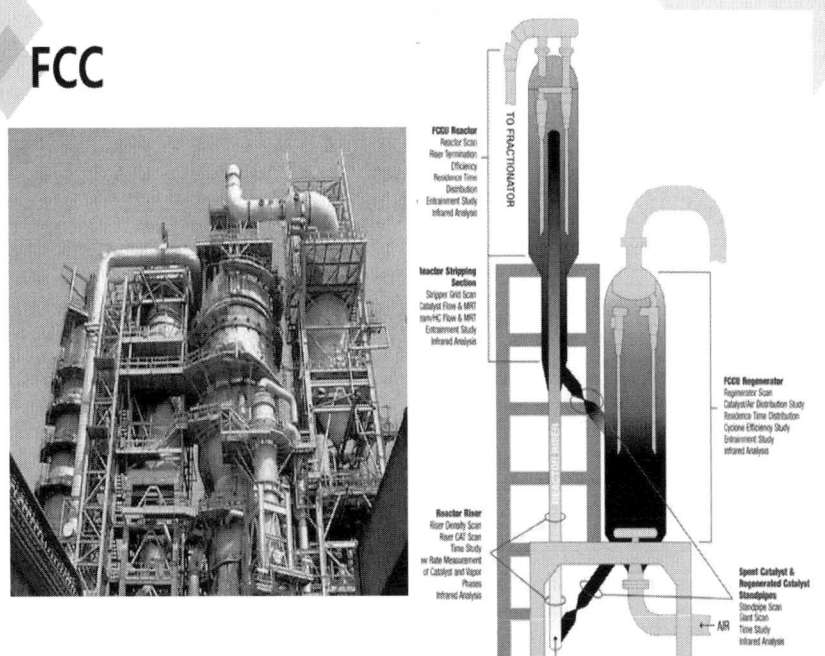

Residual Oil FCC

◈ 제8장. 화학 공정 반응기 ▶ 187

Bubble Column Reactor

> Fisher-Tropsch Reaction (Syngas to Methanol)

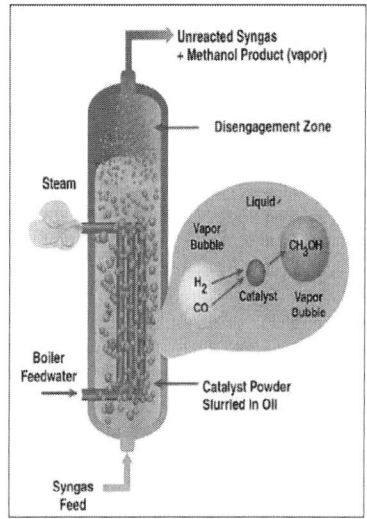

Pneumatic Reactor

> Partial Oxidation (Syngas Production)

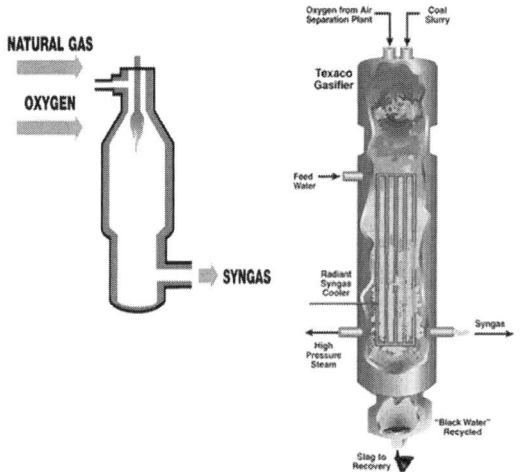

Source : globalsyngas.org/syngas-technology/syngas-production/partial-oxidation/

반응기 분류 및 특징

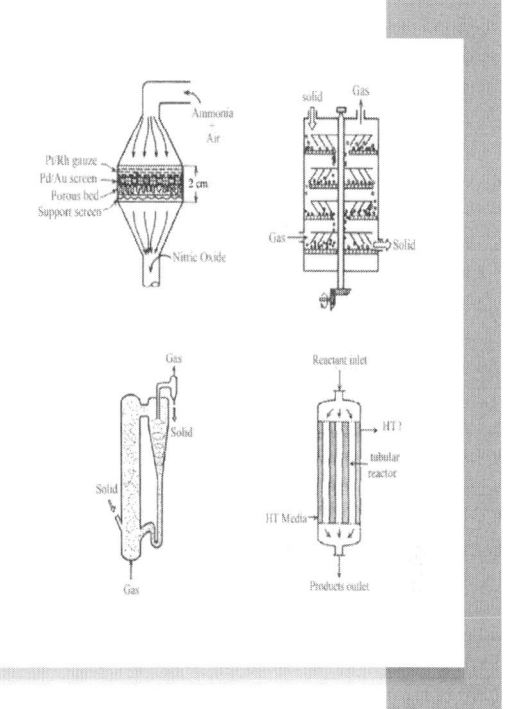

반응기 종류 및 분류

1. 고정층형 (Fixed bed)
2. 이동층형 (Moving bed)
3. 유동층형 (Fluidized bed)
4. 교반조형 (Stirred tank)
5. 기포탑형 (Bubble cap tower)
6. 관형   (Tubular)
7. 화염형   (Flammed)
8. 기류형   (Pneumatic conveying)
9. 단탑형   (Multi-staged)
10. 회전원반형 (Rotary)

고정층형 (Fixed bed)

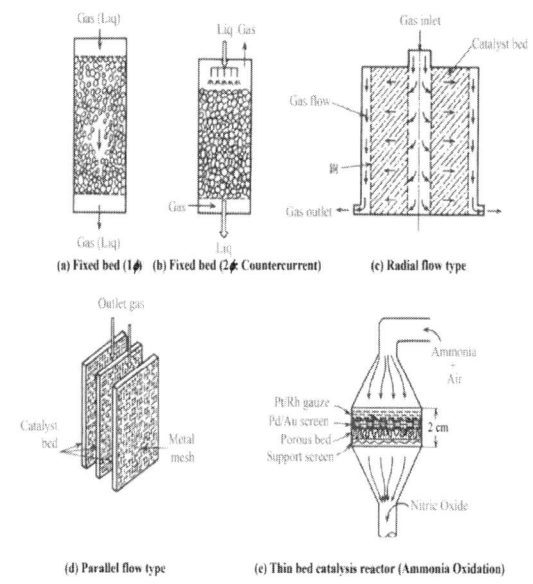

이동층형 (Moving bed)

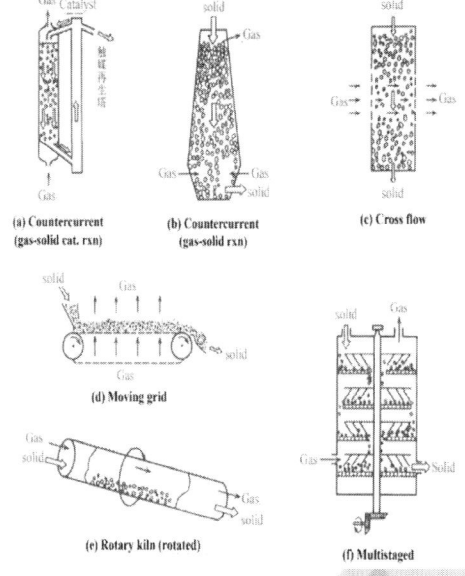

유동층형 (Fluidized bed)

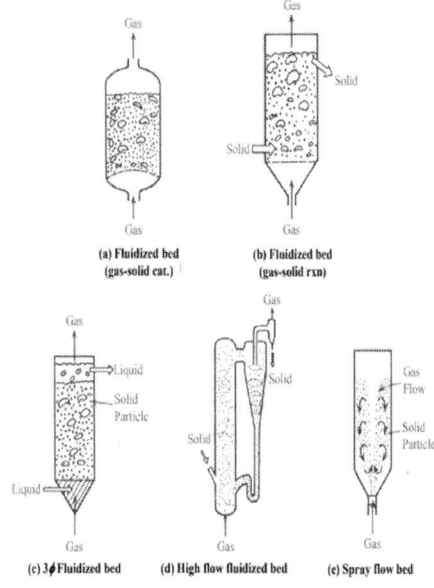

교반조형 (Stirred tank)

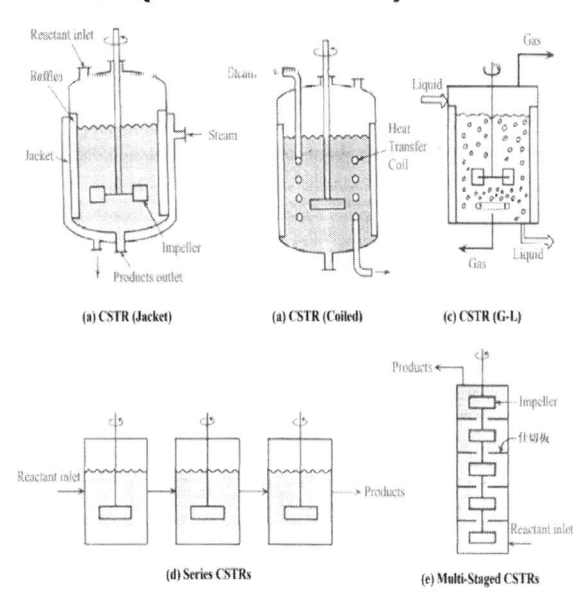

기포탑형 (Bubble cap tower)

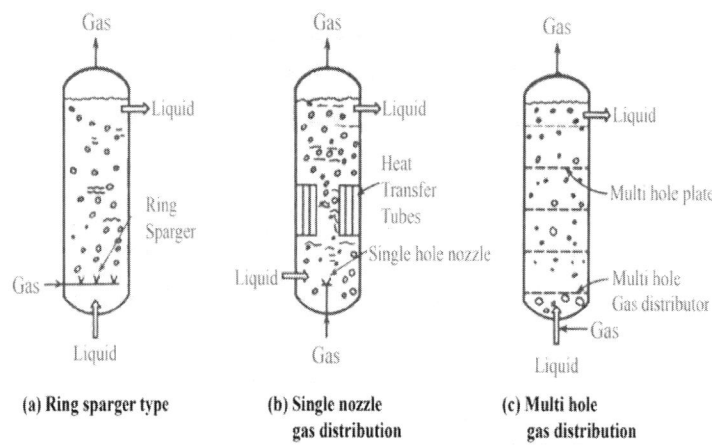

(a) Ring sparger type　　(b) Single nozzle gas distribution　　(c) Multi hole gas distribution

관형 (Tubular)

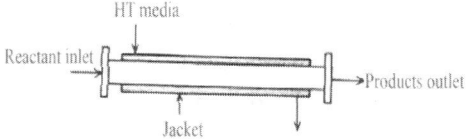

(a) Single tube type

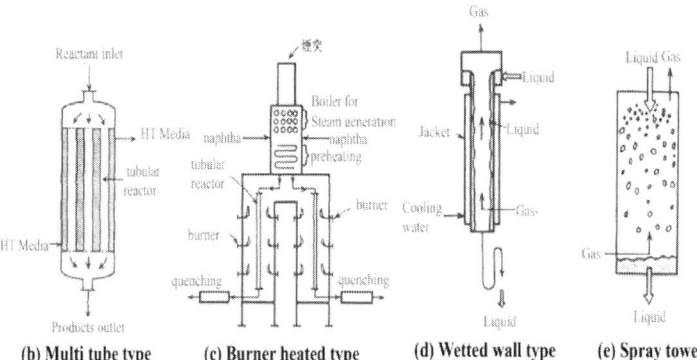

(b) Multi tube type　　(c) Burner heated type　　(d) Wetted wall type　　(e) Spray tower

단탑형 (Multi-staged)

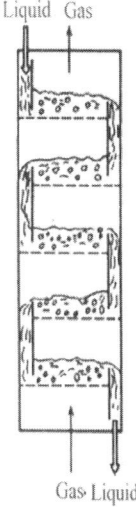

회전원반형 (Rotary)

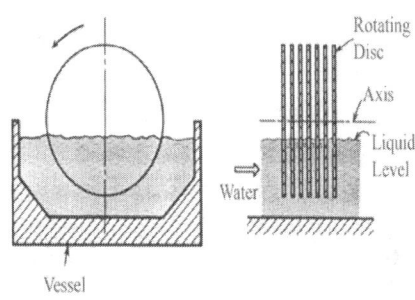

참고) Heat transfer mode (Stirred Tank)

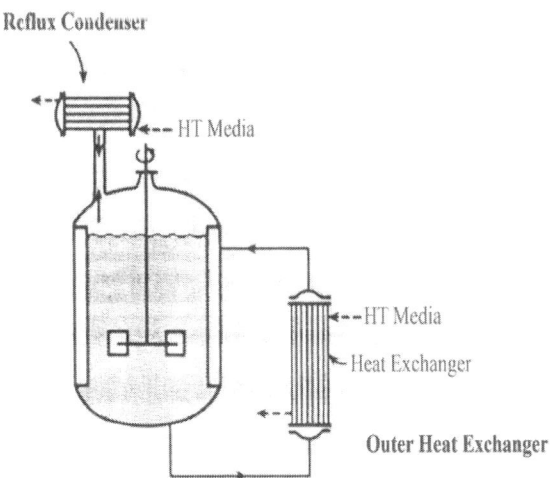

참고) Heat transfer mode (Fixed bed, catalytic)

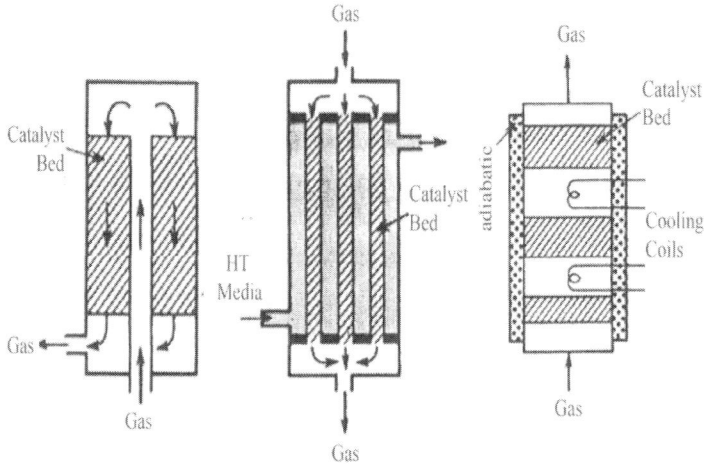

(a) Self-heat exchange (b) Multi-tube heat exchange (c) Internal cooler

Selection of Reactor Type

Reactor \ Phase	G	L	SC	GS	GL	GLS	LL	LG	SS
Fixed bed		1			2		3		
Moving bed				4					
Fluidized bed			5	6					
Stirred tank		7			8	9	10	11	
Bubble cap					12				
Tubular	13								
Pneumatic				14					

G=Gas; phase, L= Liquid phase, SC=Solid catalyst, GS=Gas-Solid phase,
GL=Gas-Liquid phase, GLS=Gas-Liquid-Solid phase, LL=Liquid-Liquid phase,
LG=Liquid-Gas phase, LS=Liquid-Solid phase, SS=Solid-Solid phase

대표적인 상업화 반응기 사례

1. Ammonia Synthesis & Naphtha Reforming Reaction
2. Hydrodesulphurization Reaction
3. Immobilized Enzyme Reaction
4. Production of Steel in Furnace
5. Production of Acrylonitrile and Fluidized Catalytic Cracking
6. Gas Phase Polymerization of Propylene
7. Polymerization of Styrene
8. Production of Antibiotics
9. Production of Terephthalic Acid & Hydrogenation of Edible Oil
10. Emulsion Polymerization of SBR
11. Production of HDPE
12. Liquid Phase Oxidation od Olefin
13. Production of Ethylene by Cracking of Naphtha
14. Production of Syngas

Batch Reactor

Characteristics	**No charge or discharge during reaction**
Phases	**Gas, Liquid, Liquid/Solid**
Application	**Small scale production** **Intermediate or one shot production** **Pharmaceutical** **Fermentation** **agricultural chemistry**
Advantages	**High conversion per unit volume for one pass** **Flexibility of operation** (same reactor can produce one product one time and a different product the next) **Easy to clean**
Disadvantages	**High operation cost** **Product quality can be changed batch to batch**

Semi-batch Reactor

Characteristics	Either one reactant is charged and the other is led continuously (at small concentrations) or else one of the product can be removed continuously to avoid side reaction.
Phases	Gas/Liquid, Liquid/Solid
Application	Small scale production Competing reactions
Advantages	High conversion per unit volume for one run Good selectivity Flexibility of operation (can be used with a reflux condenser for solvent recovery or in bubble type runs)
Disadvantages	High operation cost Product quality more variable than with continuous operation

CSTR

Characteristics	Run at steady state with continuous flow of reactants and products: the feed assumes a uniform composition through the reactor, exit stream has the same composition as in the tank
Phases	Liquid, Gas/Liquid, Liquid/Solid
Application	When agitation is required, Series configurations for different concentration streams
Advantages	Continuous operation Good temperature control Easily adapts to two phase runs Low operating (labor) cost Easy to clean
Disadvantages	Lowest conversion per unit volume By-passing and channeling possible with poor agitation

Plug Flow Reactor

Characteristics	One long reactor or many short reactors in a tube bank No radial variation in reaction rate (concentration) Changes with length down the reactor
Phases	Gas
Application	Large scale production/Continuous Production Fast reaction High Temperature
Advantages	High conversion per unit volume Low operating (labor) cost Continuous operation Good heat transfer
Disadvantages	Undesired thermal gradients Poor temperature control (hot spot) Shutdown and cleaning may be expensive

Packed-bed Reactor

Characteristics	Tubular reactor that is packed with solid catalyst
Phases	Gas/Solid catalyst, Gas/Solid
Application	Heterogeneous gas phase reaction with a catalyst
Advantages	High conversion per unit mass of catalyst Low operating (labor) cost Continuous operation
Disadvantages	Undesired thermal gradients Poor temperature control (hot spot) Channeling Shutdown and cleaning may be expensive

Fluidized-bed Reactor

Characteristics	Heterogeneous reaction
	Like a CSTR in that the reactants are well mixed
Phases	Gas/Solid catalyst, Gas/Solid
Application	Heterogeneous gas phase reaction with a catalyst
Advantages	Good mixing
	Good uniformity of temperature
	Catalyst can be continuously regenerated with the use of an auxiliary loop
Disadvantages	Bed-fluid mechanics are not well known
	Severe agitation can result in catalyst destruction and dust formation
	Uncertain scale-up

대표적인 화학공정 반응기 형태

Continuously Stirred Tank Reactor (CSTR) – complete mixing (uniform concentration everywhere)

Plug Flow Reactor (PFR) – zero axial mixing (spatially varying concentration)

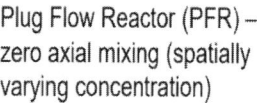
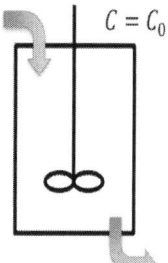
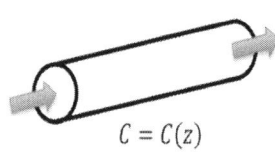

CSTR scale-up 대표적 고려 사항 (Mixing)

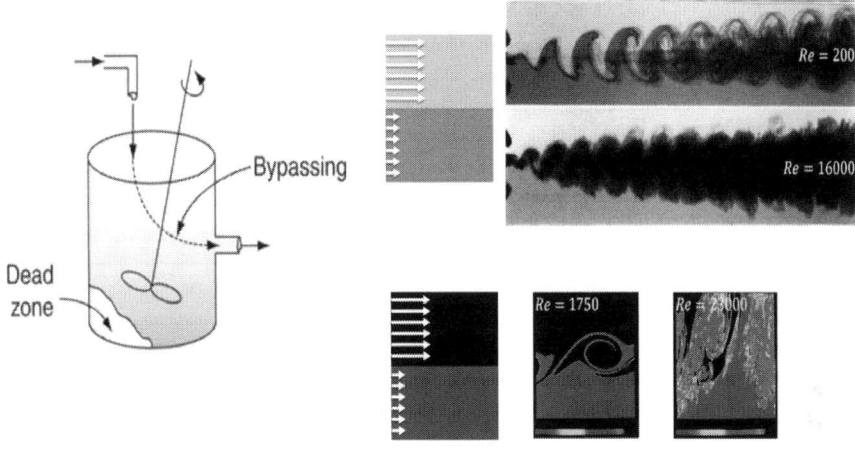

CSTR scale-up

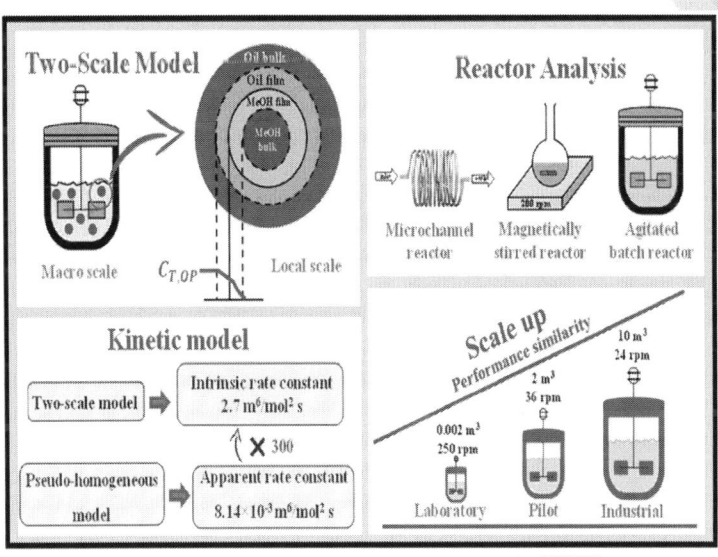

Tubular Reactors (Lab 촉매 반응기)

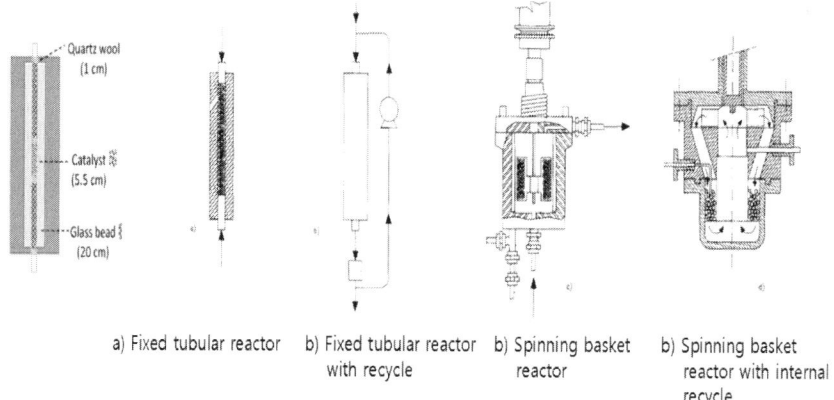

Source : Catalysis Today 34 (1997)

Tubular Reactor Scale-up (촉매 반응)

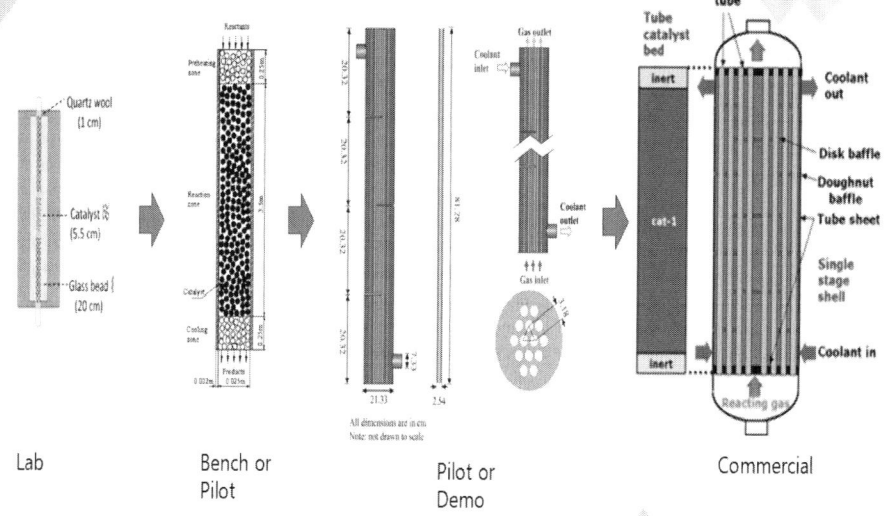

Tubular Reactor Scale-up Issue (촉매 반응)

촉매의 다양한 형상

Source : Catalyst Handbook

기공이 있는 구형 촉매에서의 Diffusion

Source : Catalysis Today 34 (1997)

반응기 재질 & 비용 산정

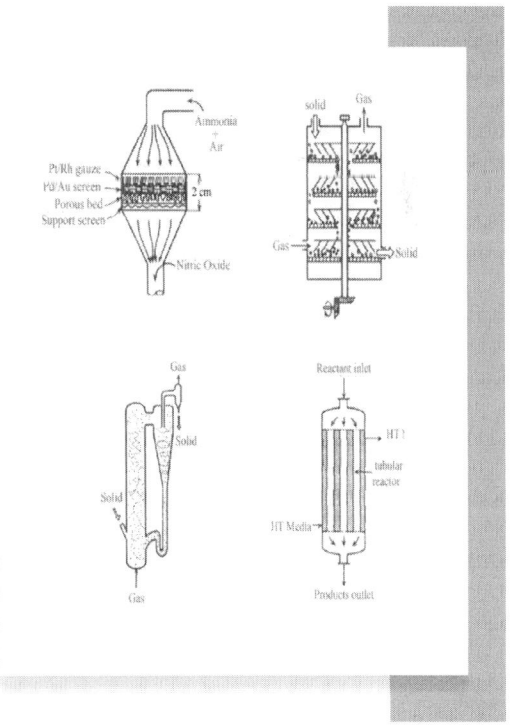

Corrosion & Erosion

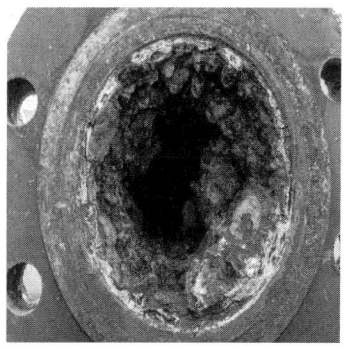 

Source : blog.envirosight.com/sewer-school-what-causes-sewer-erosion-corrosion

Corrosion Example

Source : studiousguy.com/corrosion-examples/

Corrosion (Pin-hole)

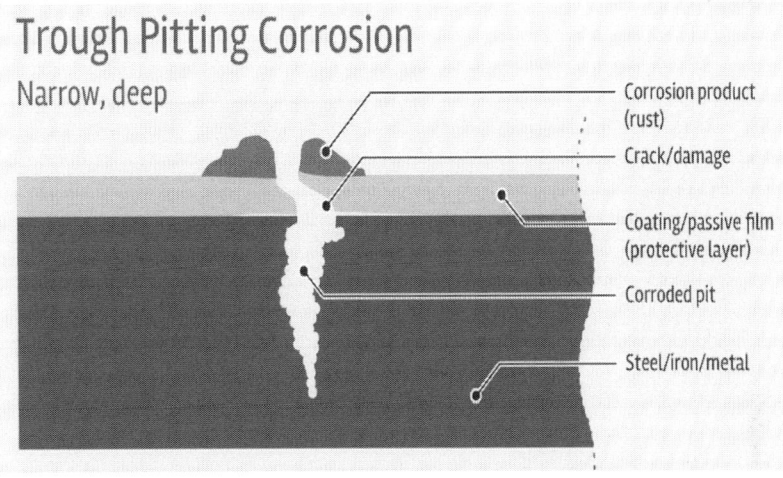

Source : www.ddcoatings.co.uk/2276/what-is-pitting-corrosion

Erosion Example

Source :
www.cdcorrosion.com/mode_corrosion/corrosion_erosion_gb.htm

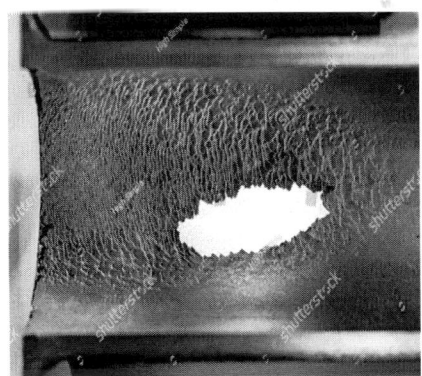

Source : www.shutterstock.com/search/erosion-o-metal

재질 선정

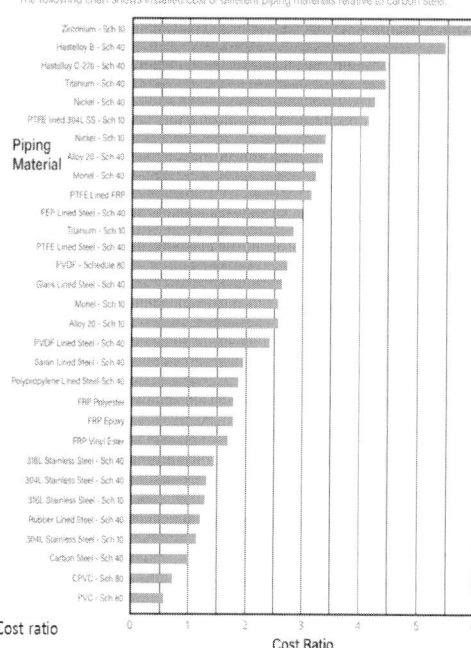

Source : Engineering ToolBox (2007), Piping materials and Cost ratio

재질 선정

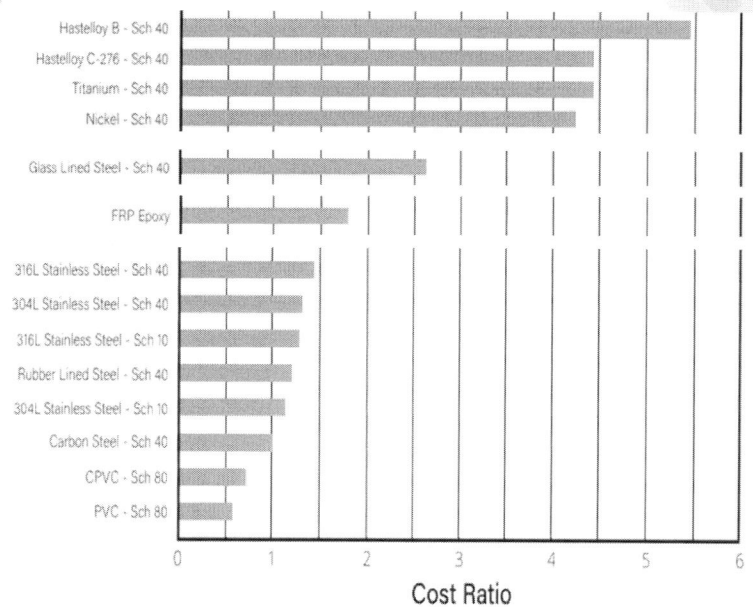

Classification of Stainless

Stainless steels
- Chromium types
  - Martensitic — Hardenable (types 403, 410, 414, 416, 416Se, 420, 431, 440A, 440B, 440C)
  - Ferritic — Nonhardenable (types 405, 430, 430F, 430Se, 442, 446)
- Chromium-nickel types
  - Austenitic — Nonhardenable, except by cold working (types 201, 202, 301, 302, 302B, 303, 303Se, 304, 304L, 305, 308, 309, 309S, 310, 310S, 314, 316, 316L, 317, 321, 347, and 348)
  - Semiaustenitic
    - Strengthened by aging or precipitation-hardening (types 17-14 CuMo, 17-10P, HNM)
    - Precipitation-hardening (PH 15-7 Mo, 17-7 PH, AM 355)
  - Martensitic — Precipitation-hardening (17-4 PH, 15-5 PH, stainless W)

참고) 주로 사용되는 Stainless Steel 예

Type	Composition, % Cr	Ni	C max.	Other significant elements	Major characteristics	Properties
301	16.00–18.00	6.00–8.00	0.15		High work-hardening rate combines cold-worked high strength with good ductility	Good structural qualities
302	17.00–19.00	8.00–10.00	0.15		Basic, general-purpose austenitic type with good corrosion resistance and mechanical properties	General-purpose
303	17.00–19.00	8.00–10.00	0.15	S 0.15 min	Free-machining modification of type 302; contains extra sulfur	Type 303Se is also available for parts involving extensive machining
304	18.00–20.00	8.00–12.00	0.08		Low-carbon variation of type 302, minimizes carbide precipitation during welding	General-purpose. Also available as 304L with 0.03% carbon to minimize carbide precipitation during welding
305	17.00–19.00	10.00–13.00	0.12		Higher heat and corrosion resistance than type 304	Good corrosion resistance
308	19.00–21.00	10.00–12.00	0.08		High Cr and Ni produce good heat and corrosion resistance. Used widely for welding rod	In order of their numbers, these alloys show increased resistance to high-temperature corrosion. Types 308S, 309S, and 310S are also available for welded construction
309	22.00–24.00	12.00–15.00	0.20		High strength and resistance to scaling at high temperatures	
310	24.00–26.00	19.00–22.00	0.25		Higher alloy content improves basic characteristics of type 309	
314	23.00–26.00	19.00–22.00	0.25	Si 1.5–3.0	High silicon content	Resistant to oxidation in air to 1100°C
316	16.00–18.00	10.00–14.00	0.08	Mo 2.00–3.00	Mo improves general corrosion and pitting resistance and high-temperature strength over that of type 302	Resistant to high pitting corrosion. Also available as 316L for welded construction

Characteristics for Temperature

> Resistance of Stainless steels to oxidation in air

Maximum temperature, °C	Stainless steel type
650	416
700	403, 405, 410, 414
800	430F
850	430, 431
900	302, 303, 304, 316, 317, 321, 347, 348, 17-14 CuMo
1000	302B, 308, 442
1100	309, 310, 314, 329, 446

> Metals and alloys for low temperature process use

ASTM specification and grade	Recommended minimum service temperature, °C
Carbon and alloy steels:	
T-1	−45
A 201, A 212, flnge or firebox quality	−45
A 203, grades A and B (2.25% Ni)	−60
A 203, grades D and E (3.50% Ni)	−100
A 353 (9% Ni)	−195
Copper alloys, silicon bronze, 70-30 brass, copper	−195
Stainless steel types 302, 304L, 304, 310, 347	−255
Aluminum alloys 5052, 5083, 5086, 5154, 5356, 5454, 5456	−255

Corrosion resistance of materials (일부분)

Code designation for corrosion resistance
- A = Acceptable, can be used successfully
- C = Caution, resistance varies widely depending on conditions: used when some corrosion is permissible
- X = Unsuitable
- Blank = Information lacking

Code designation for gasket materials
- a = Asbestos, white (compressed or woven)
- b = Asbestos, blue (compressed or woven)
- c = Asbestos (compressed and rubber-bonded)
- d = Asbestos (woven and rubber-frictioned)
- e = GR-S or natural rubber
- f = Teflon

	Metals								Nonmetals					
Chemical	Iron and steel	Cast iron (Ni-resist)	Stainless steel 18-8	18-8 Mo	Nickel	Monel	Red brass	Aluminum	Industrial glass	Carbon (Karbate)	Phenolic resins (Haveg)	Acrylic resins (Lucite)	Vinylidene chloride (Saran)	Acceptable nonmetallic gasket materials
Acetic acid, crude	C	C	C	C	C	C	C	A	A	A	A	A	C	b, c, d, f
Acetic acid, pure	X	X	C	A	C	A	X	A	A	A	A	X	X	b, c, d, f
Acetic anhydride	C	C	A	A	A	A	X	A	A	A	A	X	C	b, c, d, f
Acetone	A	A	A	A	A	A	A	A	A	A	C	X	C	a, e, f
Aluminum chloride	X	C	X	X	C	C	A	A	A	A	A	...	A	a, c, e, f
Aluminum sulfate	X	C	C	A	C	C	X	A	A	A	A	...	A	a, c, d, e, f
Alums	X	C	C	A	C	A	X	A	A	A	A	A	A	a, c, d, e, f
Ammonia (gas)	A	A	C	A	A	A	X	C	A	...	A	...	C	a, f
Ammonium chloride	C	A	C	C	A	A	C	C	A	A	A	A	A	b, c, d, e, f
Ammonium hydroxide	A	A	A	A	C	C	X	C	A	...	A	A	C	a, c, d, f
Ammonium phosphate, monobasic	X	C	A	A	...	C	X	X	A	A	A	...	...	b, c, d, e, f
Ammonium phosphate, dibasic	C	A	A	A	...	A	C	C	A	A	A	...	...	a, c, d, e, f
Ammonium phosphate, tribasic	A	A	A	A	A	A	X	C	A	A	A	...	...	a, c, d, e, f
Ammonium sulfate	C	A	C	C	A	A	C	A	A	A	A	A	A	b, c, d, e, f
Aniline	A	A	A	A	...	A	X	...	A	A	C	...	C	a, f
Benzene, benzol	A	A	A	A	A	A	A	A	A	A	A	...	C	a, f
Boric acid	X	C	A	A	A	A	C	A	A	A	A	...	A	a, c, d, e, f
Bromine	X	C	C	C	C	C	C	...	A	C	X	...	X	b, f

Corrosion resistance of materials (일부분)

Code designation for corrosion resistance
- A = Acceptable, can be used successfully
- C = Caution, resistance varies widely depending on conditions; used when some corrosion is permissible
- X = Unsuitable
- Blank = Information lacking

Code designation for gasket materials
- a = Asbestos, white (compressed or woven)
- b = Asbestos, blue (compressed or woven)
- c = Asbestos (compressed and rubber-bonded)
- d = Asbestos (woven and rubber-frictioned)
- e = GR-S or natural rubber
- f = Teflon

	Metals								Nonmetals					
Chemical	Iron and steel	Cast iron (Ni-resist)	Stainless steel 18-8	Stainless steel 18-8 Mo	Nickel	Monel	Red brass	Aluminum	Industrial glass	Carbon (Karbate)	Phenolic resins (Haveg)	Acrylic resins (Lucite)	Vinylidene chloride (Saran)	Acceptable nonmetallic gasket materials
Hydrofluoric acid	C	X	X	X	C	C	X	X	X	A	C	...	C	b, f
Nitric acid	X	C	C	C	X	X	X	C	A	C	C	...	C	b, f
Sulfuric acid, 75-95%	A	C	X	X	X	C	X	X	A	C	X	X	C	b, f
Sulfuric acid, 10-75%	X	C	X	X	C	C	X	X	A	A	C	C	A	b, f
Sulfuric acid, <10%	X	C	X	C	C	C	C	C	A	A	C	A	A	a, b, c, e, f

Installed or purchased cost of kettles

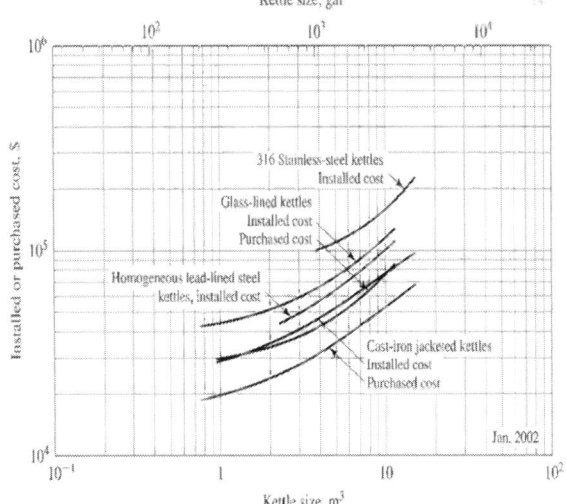

* Cost includes kettle, jacket, agitator, thermometer well, drive and support, manhole cover, and stuffing box.

◆ 제8장. 화학 공정 반응기 ▶ 207

Purchased cost of Jacketed and Stirred reactors

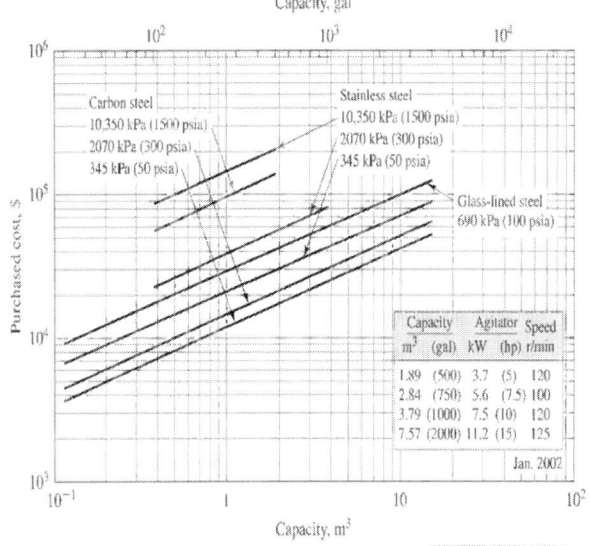

Purchased cost of Autoclave (압력용기)

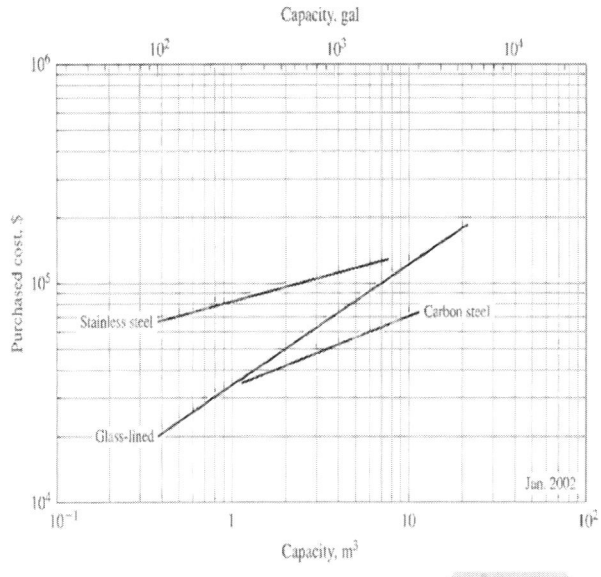

제9장
반응기 설계 1

대표적인 화학공정 반응기 형태

Continuously Stirred Tank Reactor (CSTR) – complete mixing (uniform concentration everywhere)

Plug Flow Reactor (PFR) – zero axial mixing (spatially varying concentration)

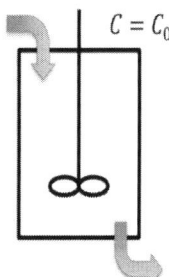

$C = C_0$

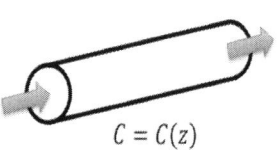

$C = C(z)$

Batch Reactor

Characteristics	No charge or discharge during reaction
Phases	Gas, Liquid, Liquid/Solid
Application	Small scale production Intermediate or one shot production Pharmaceutical Fermentation agricultural chemistry
Advantages	High conversion per unit volume for one pass Flexibility of operation (same reactor can produce one product one time and a different product the next) Easy to clean
Disadvantages	High operation cost Product quality can be changed batch to batch

Semi-batch Reactor

Characteristics	Either one reactant is charged and the other is led continuously (at small concentrations) or else one of the product can be removed continuously to avoid side reaction.
Phases	Gas/Liquid, Liquid/Solid
Application	Small scale production Competing reactions
Advantages	High conversion per unit volume for one run Good selectivity Flexibility of operation (can be used with a reflux condenser for solvent recovery or in bubble type runs)
Disadvantages	High operation cost Product quality more variable than with continuous operation

CSTR

Characteristics	Run at steady state with continuous flow of reactants and products: the feed assumes a uniform composition through the reactor, exit stream has the same composition as in the tank
Phases	Liquid, Gas/Liquid, Liquid/Solid
Application	When agitation is required, Series configurations for different concentration streams
Advantages	Continuous operation Good temperature control Easily adapts to two phase runs Low operating (labor) cost Easy to clean
Disadvantages	Lowest conversion per unit volume By-passing and channeling possible with poor agitation

Plug Flow Reactor

Characteristics	One long reactor or many short reactors in a tube bank No radial variation in reaction rate (concentration) Changes with length down the reactor
Phases	Gas
Application	Large scale production/Continuous Production Fast reaction High Temperature
Advantages	High conversion per unit volume Low operating (labor) cost Continuous operation Good heat transfer
Disadvantages	Undesired thermal gradients Poor temperature control (hot spot) Shutdown and cleaning may be expensive

Packed-bed Reactor

Characteristics	Tubular reactor that is packed with solid catalyst
Phases	Gas/Solid catalyst, Gas/Solid
Application	Heterogeneous gas phase reaction with a catalyst
Advantages	High conversion per unit mass of catalyst Low operating (labor) cost Continuous operation
Disadvantages	Undesired thermal gradients Poor temperature control (hot spot) Channeling Shutdown and cleaning may be expensive

Fluidized-bed Reactor

Characteristics Heterogeneous reaction
Like a CSTR in that the reactants are well mixed

Phases Gas/Solid catalyst, Gas/Solid

Application Heterogeneous gas phase reaction with a catalyst

Advantages Good mixing
Good uniformity of temperature
Catalyst can be continuously regenerated
with the use of an auxiliary loop

Disadvantages Bed-fluid mechanics are not well known
Severe agitation can result in catalyst destruction
and dust formation
Uncertain scale-up

Reaction design evaluation flowchart

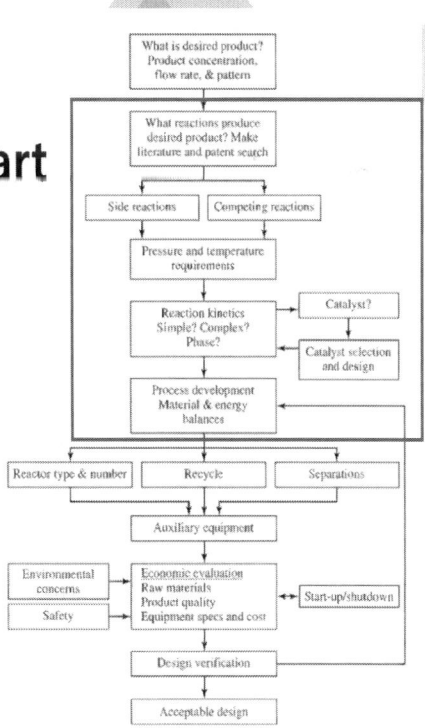

Modeling of Chemical Reactors

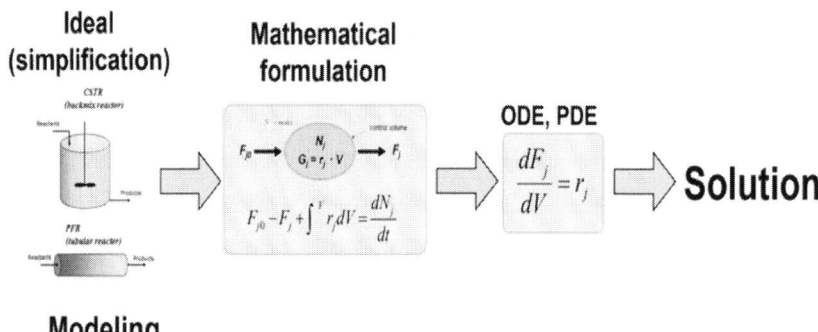

Ideal (simplification) → Mathematical formulation → ODE, PDE → Solution

Modeling

* ODE : Ordinary Differential Equation
PDE : Partial Differential Equation

Reaction rate

> Reaction rate is the rate at which a species loses its chemical identity per unit volume
> Rate of a reaction can be expressed as the rate of disappearance of a reactant or as the rate of appearance of a product

---

Consider species A:

$$A \rightarrow B$$

r_A: the rate of formation of species A per unit volume
$-r_A$: the rate of disappearance of species A per unit volume
r_B: the rate of formation of species B per unit volume

Material & Energy Balance

> Material Balance

$$\begin{pmatrix} \text{Rate of} \\ \text{accumulation of} \\ \text{species } i \text{ in the} \\ \text{volume element} \end{pmatrix} = \begin{pmatrix} \text{rate of inward} \\ \text{flow of species } i \\ \text{into the volume} \\ \text{element} \end{pmatrix} - \begin{pmatrix} \text{rate of outward} \\ \text{flow of species } i \\ \text{from the volume} \\ \text{element} \end{pmatrix} + \begin{pmatrix} \text{rate of species} \\ i \text{ generation} \\ \text{in the volume} \\ \text{element} \end{pmatrix}$$

(Control Volume)

> Energy Balance

$$\begin{pmatrix} \text{Rate of energy} \\ \text{accumulation} \\ \text{in the volume} \\ \text{element} \end{pmatrix} = \begin{pmatrix} \text{rate of inward} \\ \text{energy flow into} \\ \text{the volume} \\ \text{element} \end{pmatrix} - \begin{pmatrix} \text{rate of outward} \\ \text{energy flow from} \\ \text{the volume} \\ \text{element} \end{pmatrix} + \begin{pmatrix} \text{rate of energy} \\ \text{generation in} \\ \text{the volume} \\ \text{element} \end{pmatrix}$$

Material Balance on control volume

N_j = moles

$F_{j0} \longrightarrow \boxed{\begin{array}{c} N_j \\ G_j = r_j \cdot V \end{array}} \longrightarrow F_j$

control volume

$$G_j = \frac{moles}{time} = \frac{moles}{time \cdot volume} \cdot volume$$

A mole balance on species j, at any time, t, yields

Rate of flow of j **into** the system (mole/time)	Rate of flow of j **out of** the system (mole/time)	Rate of **generation** of j by chem. rxn within the system (mole/time)	Rate of **accumulation** of j within the system (mole/time)

in − out + generation = accumulation

$$F_{j0} - F_j + G_j = \frac{dN_j}{dt} \quad \text{Eq.(1)}$$

Rate of generation (1/2)

➢ Suppose that the of generation of species *j* for the reaction varies with the position in the *control volume*.

➢ Rate of generation

$$\Delta G_{j1} = r_{j1} \cdot \Delta V_1$$

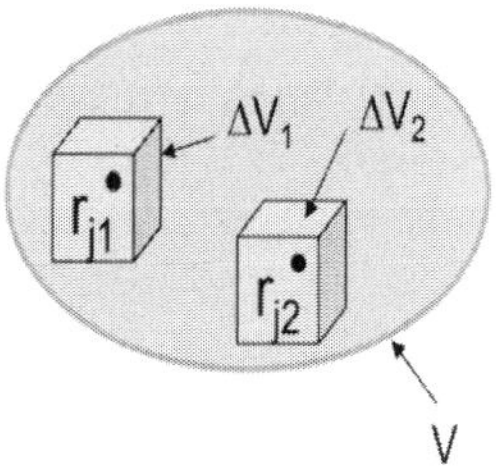

Rate of generation (2/2)

➢ If the total control volume is divided into M (sub-volume),

➢ Total rate of generation is

$$G_j = \sum_{i=1}^{M} \Delta G_{ji} = \sum_{i=1}^{M} r_{ji} \Delta V_i$$

$$\boxed{G_j = \int^V r_j \, dV} \quad \text{Eq.(2)}$$

(i.e., let M $\to \infty$ and $\Delta V \to 0$)

General Mole Balance Equation

$$F_{j0} - F_j + G_j = \frac{dN_j}{dt}$$ Eq.(1)

$$G_j = \int^V r_j \, dV$$ Eq.(2)

$$F_{j0} - F_j + \int^V r_j dV = \frac{dN_j}{dt}$$ Eq.(3)

General Mole Balance Equation (GMBE)

$$F_{j0} - F_j + \int^V r_j dV = \frac{dN_j}{dt}$$ Eq.(3)

- With **GMBE**, we can develop the design equations for the various types of reactors : batch, semi-batch, and continuous-flow

- And, upon evaluation of these equations we can determine the time (batch) or reactor volume (continuous-flow) necessary to convert a specified amount of reactants to products.

GMBE 활용

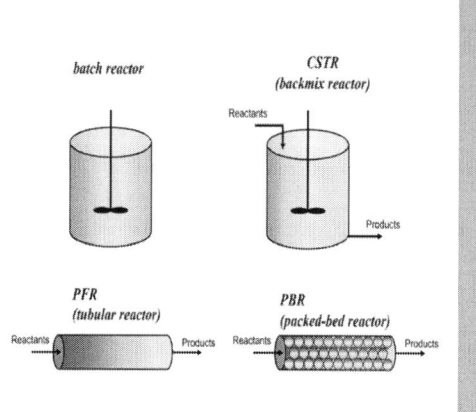

가장 일반적인 상업화 반응기 형태

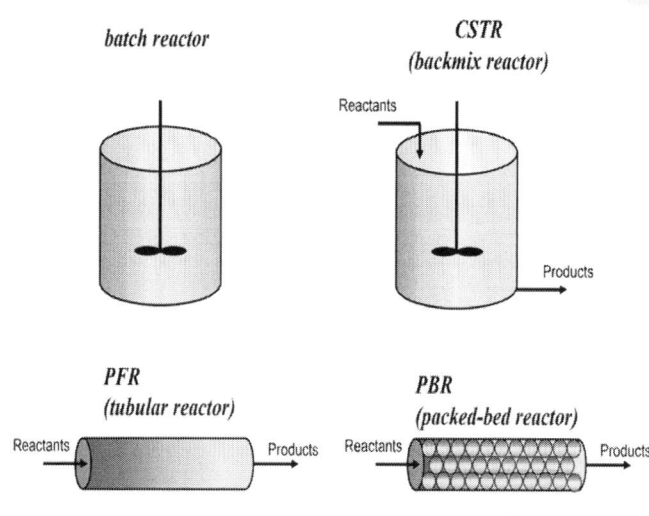

Ideal Reactor Condition

➢ Batch Reactor
- Uniform composition everywhere in the reactor
- The composition changes with time

➢ CSTR
- Uniform composition everywhere in the reactor (well mixed)
- Same composition at the reactor outlet

➢ Tubular Reactor (Plug flow reactor)
- Fluid passes through the reactor with no mixing earlier and later entering fluid, and with no overtaking
- It is as if the fluid moved in single file through the reactor
- There is no radial variation in concentration

Batch Reactor

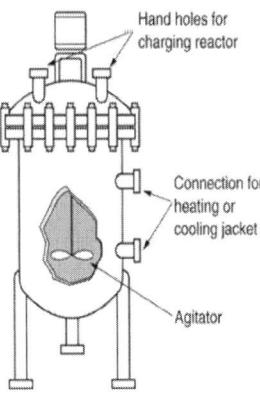 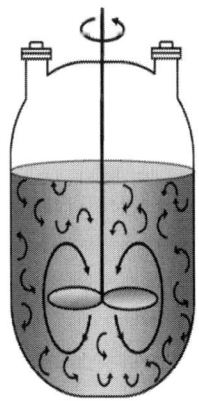

Simple batch homogeneous reactor   Batch reactor mixing pattern

Source : encyclopedai.che.engin.umich.edu/pages/reactors/batch/batch.html

Ideal Reactor Condition

> Batch Reactor
- Uniform composition everywhere in the reactor
- The composition changes with time

> CSTR
- Uniform composition everywhere in the reactor (well mixed)
- Same composition at the reactor outlet

> Tubular Reactor (Plug flow reactor)
- Fluid passes through the reactor with no mixing earlier and later entering fluid, and with no overtaking
- It is as if the fluid moved in single file through the reactor
- There is no radial variation in concentration

Batch Reactor

✓ A batch reactor has neither inflow nor outflow of reactants or products while reaction is being carried out

$$\therefore F_{j0} = F_j = 0$$

$$F_{j0} - F_j + \int^V r_j dV = \frac{dN_j}{dt} \implies \cancel{F_{j0}}^0 - \cancel{F_j}^0 + \int^V r_j dV = \frac{dN_j}{dt}$$

✓ If the reaction mixture is perfectly mixed, so that there is no variation in the rate of reaction throughout the reactor volume, we can take r_j out of integral, integrate, and write the mole balance in the form

$$\frac{dN_j}{dt} = r_j V$$

Design Equation for Batch reactor

Batch Reactor

✓ Let's consider the isomerization of species A in a batch reactor

$$A \longrightarrow B$$

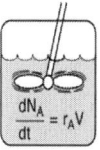
Batch Reactor

✓ As the reaction proceeds, the number of moles of A decreases and the number of moles of B increases

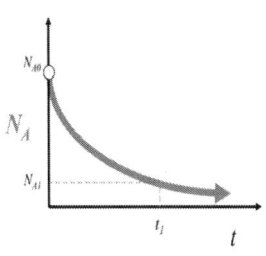
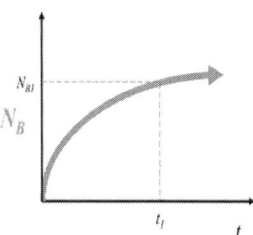

Mole-time trajectories

Batch Reactor

✓ We might ask time what a time, t_1, is necessary to reduce the initial number of moles from N_{A0} to a final desired number N_{A1}

Design Equation for Batch reactor
$$\frac{dN_j}{dt} = r_j V \implies dt = \frac{dN_A}{r_A V}$$

Integrating with limits that
at $t = 0$, $N_A = N_{A0}$
at $t = t_1$, $N_A = N_{A1}$

$$t_1 = \int_{N_{A1}}^{N_{A0}} \frac{dN_A}{-r_A V}$$

Batch Reactor (summary)

$$A \longrightarrow B$$

Design Equation for Batch reactor

$$\frac{dN_A}{dt} = r_A V$$

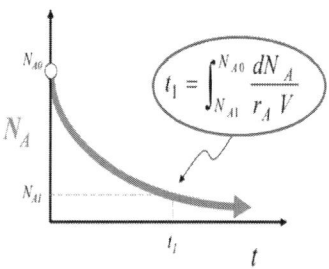

$$t_1 = \int_{N_{A1}}^{N_{A0}} \frac{dN_A}{r_A V}$$

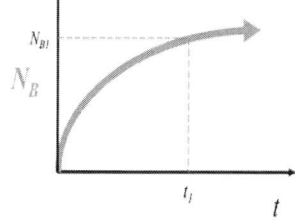

Moles of A change with time   Moles of B increase with time

CSTR

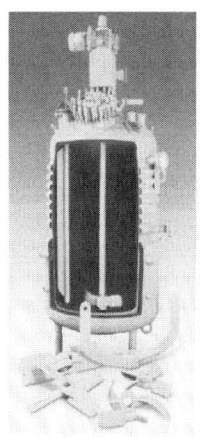

CSTR / Batch reactor

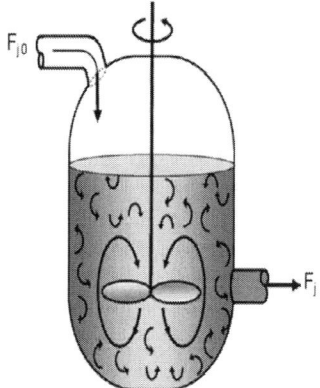

CSTR mixing pattern

Source : encyclopedai.che.engin.umich.edu/pages/reactors/CSTR/CSTR.html

Ideal Reactor Condition

➢ Batch Reactor
- Uniform composition everywhere in the reactor
- The composition changes with time

➢ CSTR
- Uniform composition everywhere in the reactor (well mixed)
- Same composition at the reactor outlet

➢ Tubular Reactor (Plug flow reactor)
- Fluid passes through the reactor with no mixing earlier and later entering fluid, and with no overtaking
- It is as if the fluid moved in single file through the reactor
- There is no radial variation in concentration

CSTR

✓ CSTR is normally run at **steady state**
 → condition do not change with time $\quad \therefore \dfrac{dN_j}{dt} = 0$

$$F_{j0} - F_j + \int^V r_j dV = \dfrac{dN_j}{dt} \quad \Longrightarrow \quad F_{j0} - F_j + \int^V r_j dV = \cancel{\dfrac{dN_j}{dt}}^{0}$$

✓ CSTR is assumed to be **perfect mixed**
 → there are no spatial variation in the rate of reaction

$$\therefore \int^V r_j \, dV = V r_j$$

CSTR

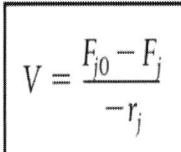

$$V = \frac{F_{j0} - F_j}{-r_j}$$

Design Equation for CSTR

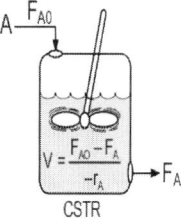

The CSTR design equation gives the reactor volume V necessary to reduce the entering molar flow rate of species j from F_{j0} to the exit molar flow rate F_j, when species j is disappearing at a rate of $-r_j$. We note that the CSTR is modeled such that the conditions in the exit stream (e.g., concentration and temperature) **are identical** to those in the tank. The molar flow rate F_j is just the product of the concentration of species j and the volumetric flow rate v

$$F_j = C_j \cdot v$$
$$\frac{\text{moles}}{\text{time}} = \frac{\text{moles}}{\text{volume}} \cdot \frac{\text{volume}}{\text{time}}$$

CSTR

$$V = \frac{F_{j0} - F_j}{-r_j} \quad \Longleftrightarrow \quad V = \frac{v_0 C_{A0} - v C_A}{-r_A}$$

Design Equation for CSTR

Tubular reactor (Plug Flow)

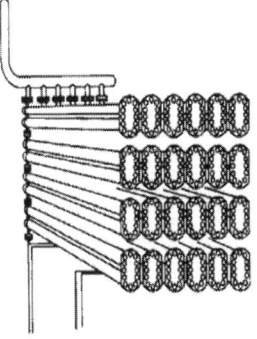

Tubular reactor schematic.

Tubular reactor for Dimersol G

Source : encyclopedai.che.engin.umich.edu/pages/reactors/PFR/PFR.html

Ideal Reactor Condition

➢ Batch Reactor
- Uniform composition everywhere in the reactor
- The composition changes with time

➢ CSTR
- Uniform composition everywhere in the reactor (well mixed)
- Same composition at the reactor outlet

➢ Tubular Reactor (Plug flow reactor)
- Fluid passes through the reactor with no mixing earlier and later entering fluid, and with no overtaking
- It is as if the fluid moved in single file through the reactor
- There is no radial variation in concentration

Tubular reactor (Plug Flow)

✓ In the tubular reactor, the reactants are continually consumed as they flow down the length of the reactor

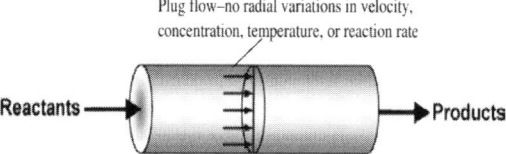

✓ The concentration varies continuously in the axial direction through the reactor

✓ Consequently, the reaction rate will also vary axially

Tubular reactor (Plug Flow)

✓ To develop the PFR design equation, we shall divide (conceptually) the reactor into a number of sub-volumes so that within each sub-volume ΔV, the reaction rate may be considered spatially uniform.

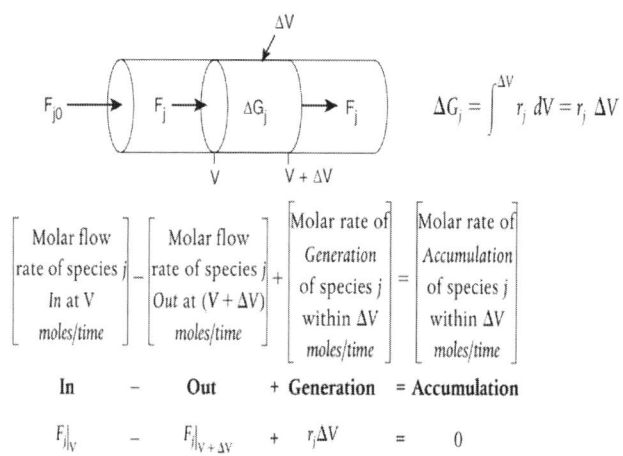

$$\Delta G_j = \int^{\Delta V} r_j \, dV = r_j \, \Delta V$$

$$\begin{bmatrix} \text{Molar flow} \\ \text{rate of species } j \\ \text{In at } V \\ \text{moles/time} \end{bmatrix} - \begin{bmatrix} \text{Molar flow} \\ \text{rate of species } j \\ \text{Out at } (V+\Delta V) \\ \text{moles/time} \end{bmatrix} + \begin{bmatrix} \text{Molar rate of} \\ \text{Generation} \\ \text{of species } j \\ \text{within } \Delta V \\ \text{moles/time} \end{bmatrix} = \begin{bmatrix} \text{Molar rate of} \\ \text{Accumulation} \\ \text{of species } j \\ \text{within } \Delta V \\ \text{moles/time} \end{bmatrix}$$

In − Out + Generation = Accumulation

$$F_j\big|_V \quad - \quad F_j\big|_{V+\Delta V} \quad + \quad r_j \Delta V \quad = \quad 0$$

Tubular reactor (Plug Flow)

$$F_j|_V - F_j|_{V+\Delta V} + r_j \Delta V = 0 \implies \left[\frac{F_j|_{V+\Delta V} - F_j|_V}{\Delta V}\right] = r_j$$

Design Equation for PFR
$$\boxed{\frac{dF_j}{dV} = r_j} \impliedby \lim_{\Delta V \to 0}\left[\frac{F_j|_{V+\Delta V} - F_j|_V}{\Delta V}\right] = r_j$$

✓ For species A, the mole balance is

$$\boxed{\frac{dF_A}{dV} = r_A}$$

Tubular reactor

Tubular reactor (Plug Flow)

✓ Let's consider the isomerization of species A in a PFR

$$A \longrightarrow B$$

✓ As the reactants proceed down the reactor,
A is consumed by reaction and B is produced.

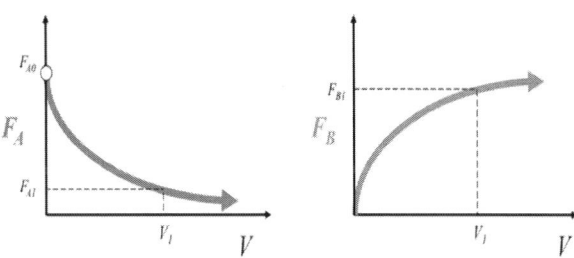

Profiles of molar flow rates in a PFR

Tubular reactor (Plug Flow)

✓ We now ask, "What is the reactor volume V_1 necessary to reduce the entering molar flow rate of A from F_{A0} to F_{A1}?

$$\frac{dF_A}{dV} = r_A \implies dV = \frac{dF_A}{r_A}$$

✓ And then, integrating with limits

$$V_1 = \int_{F_{A0}}^{F_{A1}} \frac{dF_A}{r_A} = \int_{F_{A1}}^{F_{A0}} \frac{dF_A}{-r_A}$$

at $V = 0$, then $F_A = F_{A0}$, and at $V = V_1$, then $F_A = F_{A1}$

Tubular reactor (Plug Flow) (summary)

$$A \rightarrow B$$

Design Equation for PFR
$$\frac{dF_A}{dV} = r_A$$

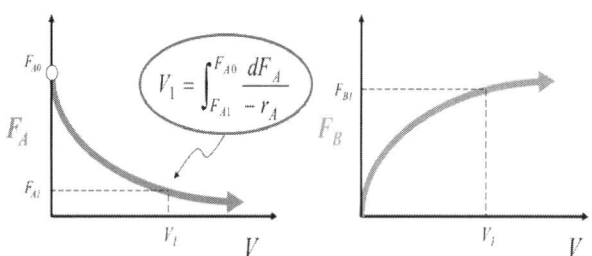

Profiles of molar flow rates in a PFR

Packed Bed Reactor (PBR)

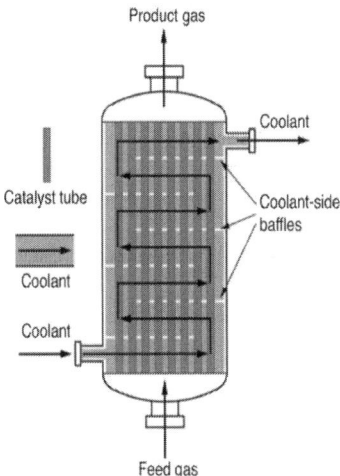

Catalytic packed-bed reactor

Source : encyclopedai.che.engin.umich.edu/pages/reactors/PBR/PBR.html

참고) Catalytic Reaction Steps

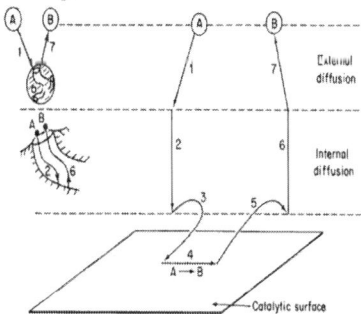

1. Mass transfer (diffusion) of the reactant(s) (e.g., species A) from the bulk fluid to the external surface of the catalyst pellet
2. Diffusion of the reactant from the pore mouth through the catalyst pores to the immediate vicinity of the internal catalytic surface
3. Adsorption of reactant A onto the catalyst surface
4. Reaction on the surface of the catalyst (e.g., A $\longrightarrow$ B)
5. Desorption of the products (e.g., B) from the surface
6. Diffusion of the products from the interior of the pellet to the pore mouth at the external surface
7. Mass transfer of the products from the external pellet surface to the bulk fluid

Packed Bed Reactor (PBR)

- ✓ The reaction takes place on the surface of the catalyst. Consequently, the reaction rate is based on mass of solid catalyst rather than on reactor volume

- ✓ For a fluid-solid heterogeneous system, the rate of reaction of a species A,

$$-r'_A = \text{mol A reacted}/(\text{time} \times \text{mass of catalyst})$$

- ✓ And then, by multiplying the heterogeneous rate, $-r'_A$, and bulk catalyst density, $\rho_b \left(\frac{mass}{volume}\right)$, we can obtain the homogeneous reaction rate $-r_A$

$$-r_A = \rho_b (-r'_A)$$

$$\left(\frac{mol}{dm^3 \cdot s}\right) = \left(\frac{g}{dm^3}\right)\left(\frac{mol}{g \cdot s}\right)$$

Packed Bed Reactor (PBR)

- ✓ The derivation of the design equation for a packed-bed catalytic reactor will be carried out in a manner analogous to the development of the tubular design equation

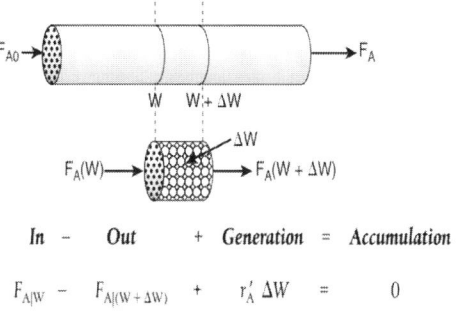

$$\begin{array}{ccccccc} In & - & Out & + & Generation & = & Accumulation \\ F_{A|W} & - & F_{A|(W+\Delta W)} & + & r'_A \Delta W & = & 0 \end{array}$$

- ✓ As with the PFR, the PBR is assumed to have no radial gradients in concentration, temperature, or reaction rate

Packed Bed Reactor (PBR)

✓ Consider the dimension of generation term

$$(r'_A)\Delta W \equiv \frac{\text{moles } A}{(\text{time})(\text{mass of catalyst})} \cdot (\text{mass of catalyst}) \equiv \frac{\text{moles } A}{\text{time}}$$

✓ As expected, the same dimension of molar flow rate F_A.

$$F_{A|W} - F_{A|(W+\Delta W)} + r'_A \Delta W = 0$$

taking the limit as $\Delta W \to 0$ | No pressure drop
No catalyst decay

Design Equation for PBR

$$\boxed{\frac{dF_A}{dW} = r'_A} \quad \text{and} \quad W = \int_{F_{A0}}^{F_A} \frac{dF_A}{r'_A} = \int_{F_A}^{F_{A0}} \frac{dF_A}{-r'_A}$$

Summary for GMBE

➤ A mole balance on species j, which enters, leaves, reacts, and accumulates in a system volume V,

$$F_{j0} - F_j + \int^V r_j dV = \frac{dN_j}{dt}$$

➤ *If, and only if,* the contents of the reactor are well mixed will the mole balance on species A,

$$F_{A0} - F_A + r_A V = \frac{dN_A}{dt}$$

Summary for Reactor Mole Balances

Reactor	Differential	Algebraic	Integral	
Batch	$\dfrac{dN_A}{dt} = r_A V$		$t = \displaystyle\int_{N_{A0}}^{N_A} \dfrac{dN_A}{r_A V}$	N_A vs t
CSTR		$V = \dfrac{F_{A0} - F_A}{-r_A}$		
PFR	$\dfrac{dF_A}{dV} = r_A$		$V = \displaystyle\int_{F_{A0}}^{F_A} \dfrac{dF_A}{r_A}$	F_A vs V
PBR	$\dfrac{dF_A}{dW} = r'_A$		$W = \displaystyle\int_{F_{A0}}^{F_A} \dfrac{dF_A}{r'_A}$	F_A vs W

Reactor Sizing

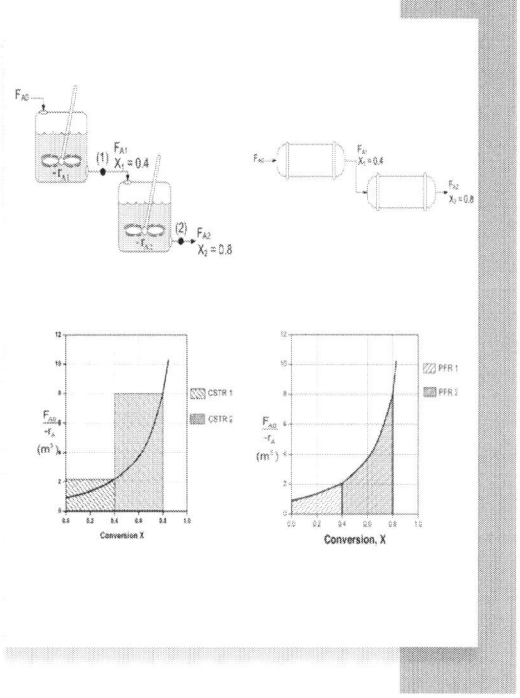

Definition of Conversion

> Consider the general reaction

$$aA + bB \longrightarrow cC + dD \implies A + \frac{b}{a}B \longrightarrow \frac{c}{a}C + \frac{d}{a}D$$

❖ Question : "How can we quantify how far a reaction proceeds to the right?" or
"How many moles of C are formed from every mole of A consumed?"

❖ Answer : A convenient way to answer these questions is to define a parameter called *conversion*.

✓ The conversion X_A is the number of moles of A that have reacted per mole of A fed to the system

$$X_A = \frac{\text{Moles of A reacted}}{\text{Moles of A fed}}$$

Batch reactor design eq.

$$[\text{Moles of A reacted (consumed)}] = [\text{Moles of A fed}] \cdot \left[\frac{\text{Moles of A reacted}}{\text{Moles of A fed}}\right]$$

$$\begin{bmatrix}\text{Moles of A} \\ \text{reacted} \\ \text{(consumed)}\end{bmatrix} = [N_{A0}] \cdot [X]$$

> Consider the number of moles of A that remain in the reactor after time t

$$\begin{bmatrix}\text{Moles of A} \\ \text{in reactor} \\ \text{at time t}\end{bmatrix} = \begin{bmatrix}\text{Moles of A} \\ \text{initially fed} \\ \text{to reactor at} \\ t=0\end{bmatrix} - \begin{bmatrix}\text{Moles of A that} \\ \text{have been consumed by chemical} \\ \text{reaction}\end{bmatrix}$$

$$[N_A] = [N_{A0}] - [N_{A0}X]$$

$$\boxed{N_A = N_{A0} - N_{A0}X = N_{A0}(1-X)}$$

Batch reactor design eq.

$$N_A = N_{A0} - N_{A0}X = N_{A0}(1-X) \implies \frac{dN_A}{dt} = 0 - N_{A0}\frac{dX}{dt}$$

$$\therefore \frac{dN_A}{dt} = r_A V \quad \text{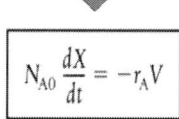 Design Equation for Batch reactor}$$

$$N_{A0}\frac{dX}{dt} = -r_A V$$

✓ This equation is called to the differential form of the design equation for batch reactor in terms of conversion

Mole Balance terms of Conversion

	Differential Form	Algebraic Form	Integral Form
Batch	$N_{A0}\dfrac{dX}{dt} = -r_A V$		$t = N_{A0}\displaystyle\int_0^X \dfrac{dX}{-r_A V}$
CSTR		$V = \dfrac{F_{A0}(X_{out} - X_{in})}{(-r_A)_{out}}$	
PFR	$F_{A0}\dfrac{dX}{dV} = -r_A$		$V = F_{A0}\displaystyle\int_{X_{in}}^{X_{out}} \dfrac{dX}{-r_A}$
PBR	$F_{A0}\dfrac{dX}{dW} = -r'_A$		$W = F_{A0}\displaystyle\int_{X_{in}}^{X_{out}} \dfrac{dX}{-r'_A}$

❖ If we know $-r_A$ as a function of X, we can size any isothermal reaction system

Reactor Sizing Example

> Consider the irreversible isothermal gas phase isomerization

$$A \longrightarrow B$$

✓ Reaction condition : 500 K, 830 kPa (8.2 atm), $F_{A0} = 0.4$ mol/s.

Table 1) Raw data

X	0	0.1	0.2	0.4	0.6	0.7	0.8
$-r_A$ (mol/m$^3 \cdot$ s)	0.45	0.37	0.30	0.195	0.113	0.079	0.05

Reactor Sizing Example

> For CSTR

$$V = \frac{F_{A0} X}{(-r_A)_{exit}}$$

> For PFR

$$V = F_{A0} \int_0^X \frac{dX}{-r_A}$$

Table 2) Processed data

X	0.0	0.1	0.2	0.4	0.6	0.7	0.8
$-r_A \left(\dfrac{mol}{m^3 \cdot s}\right)$	0.45	0.37	0.30	0.195	0.113	0.079	0.05
$(1/-r_A)\left(\dfrac{m^3 \cdot s}{mol}\right)$	2.22	2.70	3.33	5.13	8.85	12.7	20
$(F_{A0}/-r_A)(m^3)$	0.89	1.08	1.33	2.05	3.54	5.06	8.0

Reactor Sizing Example

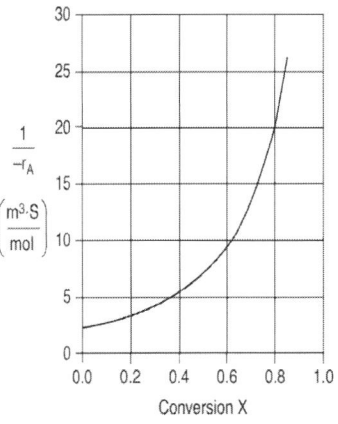

Fig 1) Processed data 1

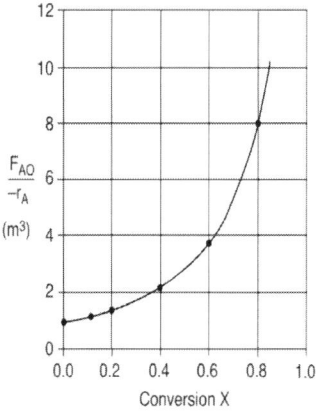

Fig 2) Processed data 2
(Levenspiel plot)

Reactor Sizing Example

> For CSTR

- Q-1) Calculate the volume necessary to achieve 80% conversion
- Q-2) Shade the area in Figure 2 that would gives the CSTR volume to achieve 80% conversion

❖ Q-1)

$$V = \frac{F_{A0}X}{(-r_A)_{exit}}$$

$$V = 0.4 \frac{mol}{s}\left(\frac{20\ m^3 \cdot s}{mol}\right)(0.8) = 6.4\ m^3$$

$$V = 6.4\ m^3 = 6400\ dm^3 = 6400\ liters$$

❖ Q-2)

$$V = \left[\frac{F_{A0}}{-r_A}\right]X$$

$$V = [8\ m^3][0.8]$$
$$= 6.4\ m^3$$
$$= 6400\ L$$

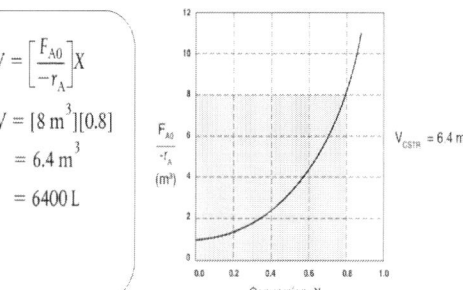

Fig 3) Levenspiel plot for Q-2

Reactor Sizing Example

> For PFR
  - Q-3) Calculate the volume necessary to achieve 80% conversion
  - Q-4) Shade the area in Figure 2 that would gives the CSTR volume to achieve 80% conversion

❖ Q-3)

$$F_{A0}\frac{dX}{dV} = -r_A$$

$$V = F_{A0}\int_0^{0.8} \frac{dX}{-r_A} = \int_0^{0.8} \frac{F_{A0}}{-r_A}dX$$

$$V = 2.165 \text{ m}^3 = 2165 \text{ dm}^3$$

* V can calculate using the five-point quadarature formular

❖ Q-4)

$$V = \int_0^{0.8} \frac{F_{A0}}{-r_A}dX$$

= Area under the curve between $X = 0$ and $X = 0.8$

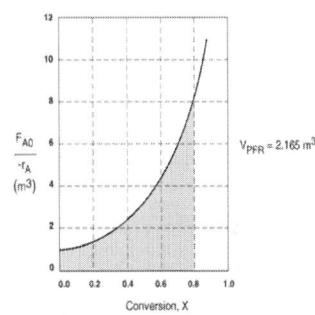

Fig 4) Levenspiel plot for Q-4

Comparing CSTR and PFR size

X	0	0.1	0.2	0.4	0.6	0.7	0.8
$-r_A$ (mol/m³·s)	0.45	0.37	0.30	0.195	0.113	0.079	0.05

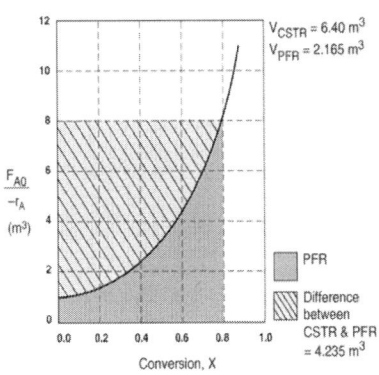

Fig 5) Comparison of CSTR and PFR reactor size for X=0.8

CSTR in series

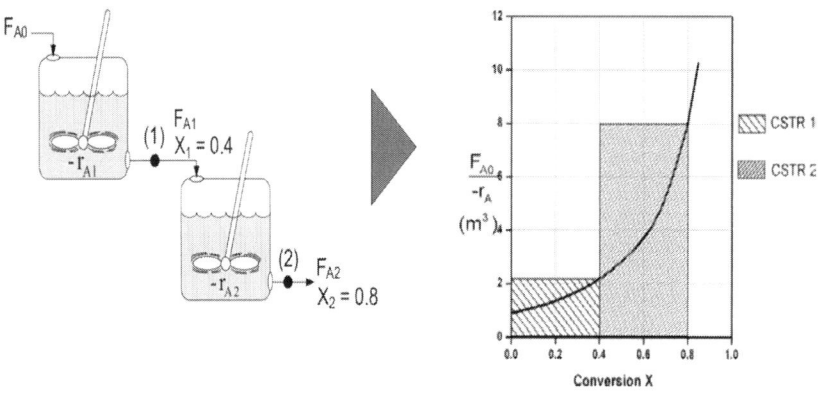

Fig 6) Levenspiel plot for CSTRs in series

Approximating a PFR by CSTR

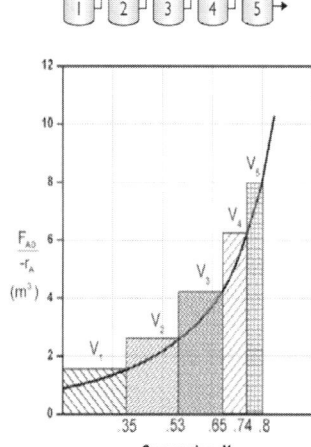

❖ The fact that we can model a PFR with a large number of CSTRs is an important result

Fig 7) Levenspiel plot showing comparison of CSTRs in series with one PFR

PFR in series

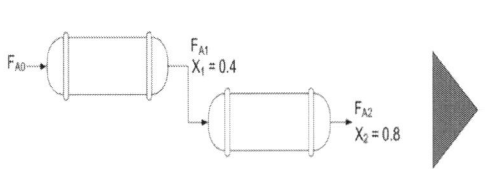

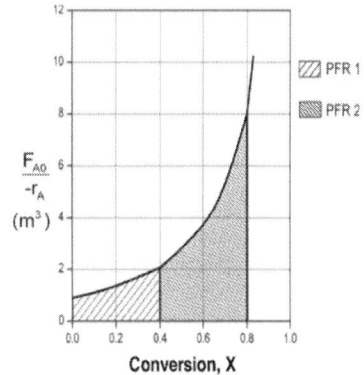

❖ The overall conversion of two PFRs in series is the same as one PFR with the same total volume

Fig 8) Levenspiel plot for two PFRs in series

Combination of CSTRs and PFRs

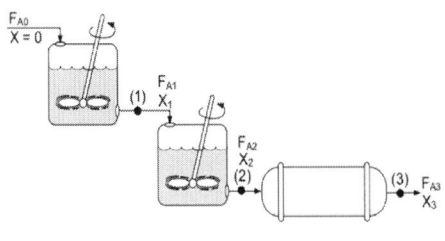

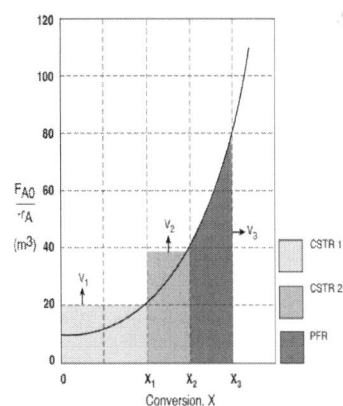

Combination of CSTRs and PFRs

X	0.0	0.2	0.4	0.6	0.65
$-r_A$ (kmol/m³·h)	39	53	59	38	25

⬇

X	0.0	0.2	0.4	0.6	0.65
$-r_A$ (kmol/m³·h)	39	53	59	38	25
$[F_{A0}/-r_A]$ (m³)	1.28	0.94	0.85	1.32	2.0

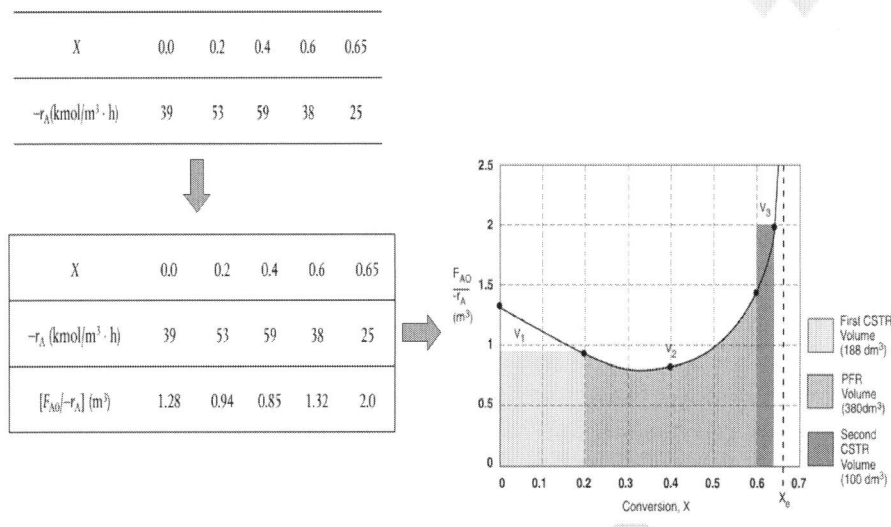

First CSTR Volume (188 dm³)
PFR Volume (380 dm³)
Second CSTR Volume (100 dm³)

Summary for Mole Balances (conversion)

	Differential Form	Algebraic Form	Integral Form
Batch	$N_{A0}\dfrac{dX}{dt} = -r_A V$		$t = N_{A0}\displaystyle\int_0^X \dfrac{dX}{-r_A V}$
CSTR		$V = \dfrac{F_{A0}(X_{out} - X_{in})}{(-r_A)_{out}}$	
PFR	$F_{A0}\dfrac{dX}{dV} = -r_A$		$V = F_{A0}\displaystyle\int_{X_{in}}^{X_{out}} \dfrac{dX}{-r_A}$
PBR	$F_{A0}\dfrac{dX}{dW} = -r'_A$		$W = F_{A0}\displaystyle\int_{X_{in}}^{X_{out}} \dfrac{dX}{-r'_A}$

Graphical Integration (Levenspiel plots)

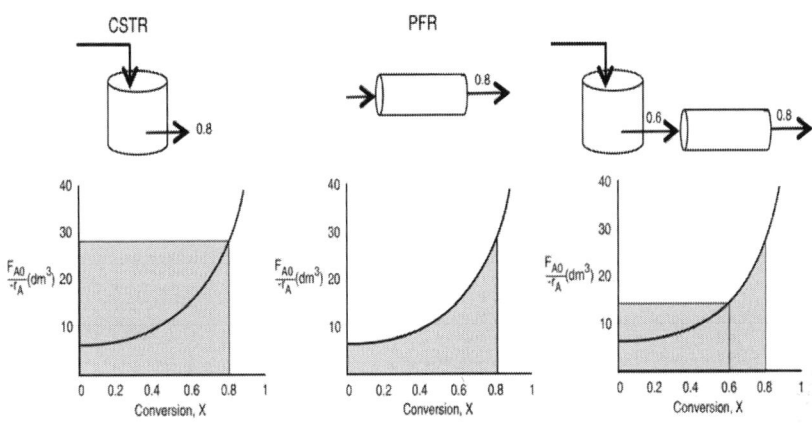

강 의 노 트

🚗 1.

🚗 2.

🚗 3.

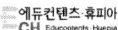

제10장
반응기 설계 2

반응기 Scale-up

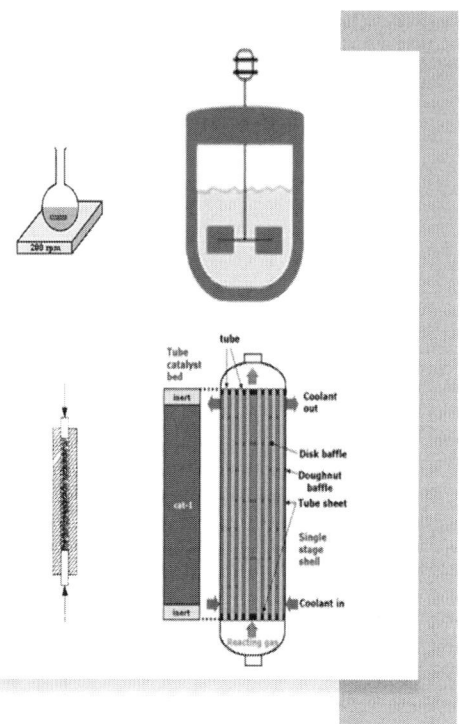

CSTR scale-up 대표적 고려 사항 (Mixing)

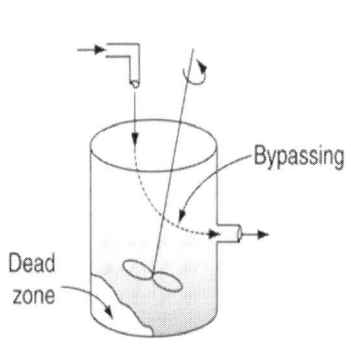

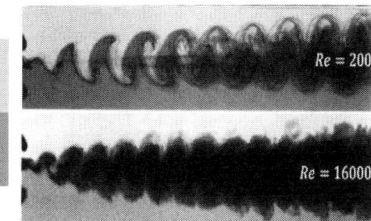

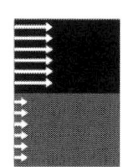

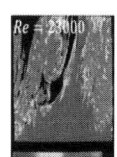

CSTR scale-up

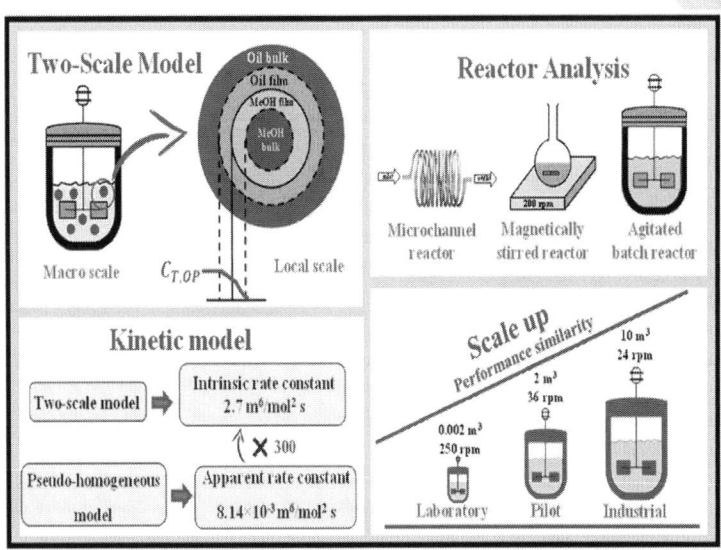

CSTR CFD example (with & w/o Baffle)

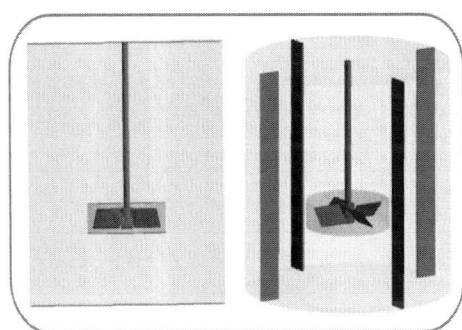

Geometry

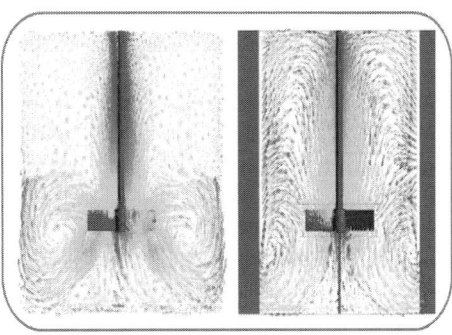

Velocity vector

CSTR CFD example (with & w/o Baffle)

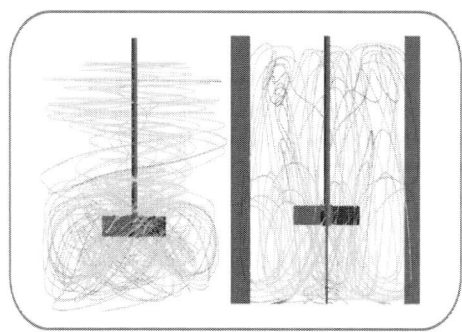
Streamline

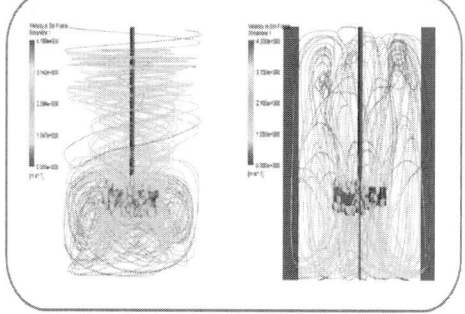
Streamline w/ RPM

CSTR CFD example (turbine impeller with Baffle)

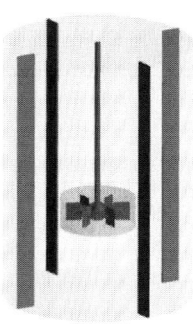
Geometry

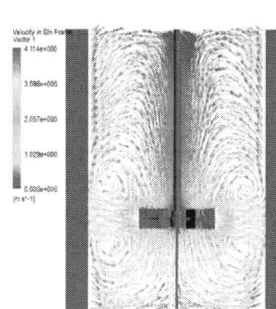
Velocity vector

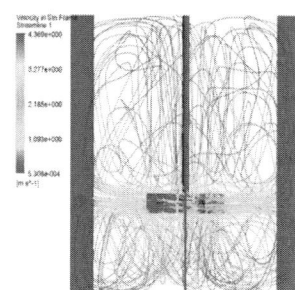
Streamline

Kinetic model example (Hydroformylation)

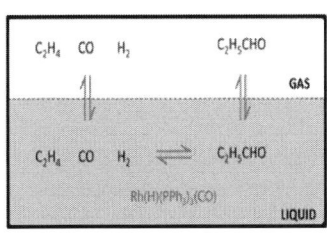

Kinetic model example (Hydroformylation)

$$r_1 = k_{1,+}\left(n_{Rh(CO)_2(H)(L)} \cdot C_{C_2H_4} - \frac{1}{K_1} \cdot n_{Rh(CO)_2(C_2H_4)(H)(L)}\right) [mol\ s^{-1}]$$

$$r_5 = k_{5,+}\left(n_{Rh(CO)_2(COC_2H_5)(L)} \cdot C_{H_2} - \frac{1}{K_5} \cdot n_{Rh(CO)_2(COC_2H_5)(H)_2(L)}\right) [mol\ s^{-1}]$$

$$r_2 = k_{2,+}\left(n_{Rh(CO)_2(C_2H_4)(L)} - \frac{1}{K_2} \cdot n_{Rh(CO)_2(C_2H_5)(L)}\right) [mol\ s^{-1}]$$

$$r_6 = k_{6,+}\left(n_{Rh(CO)_2(H)(L)} \cdot C_{propanal} - \frac{1}{K_6} \cdot n_{Rh(CO)_2(COC_2H_5)(H)_2(L)}\right) [mol\ s^{-1}]$$

$$r_3 = k_{3,+}\left(n_{Rh(CO)_2(C_2H_5)(L)} \cdot C_{CO} - \frac{1}{K_3} \cdot n_{Rh(CO)_3(C_2H_5)(L)}\right) [mol\ s^{-1}]$$

$$r_7 = k_{3,+}\left(n_{Rh(CO)_2(H)(L)} \cdot C_{CO} - \frac{1}{K_3'} \cdot n_{Rh(CO)_3(H)(L)}\right) [mol\ s^{-1}]$$

$$r_4 = k_{4,+}\left(n_{Rh(CO)_3(C_2H_5)(L)} - \frac{1}{K_4} \cdot n_{Rh(CO)_2(COC_2H_5)(L)}\right) [mol\ s^{-1}]$$

$$r_8 = k_{3,+}\left(n_{Rh(CO)_2(COC_2H_5)(L)} \cdot C_{CO} - \frac{1}{K_3} \cdot n_{Rh(CO)_3(COC_2H_5)(L)}\right) [mol\ s^{-1}]$$

$$r = k\left(\frac{C_{C_2H_4}}{C^\circ}\right)^a \left(\frac{C_{CO}}{C^\circ}\right)^b \left(\frac{C_{H_2}}{C^\circ}\right)^c \left(\frac{C_{Rh}}{C^\circ}\right)^d [mol\ m^{-3}\ s^{-1}]$$

Kinetic model example (Hydroformylation)

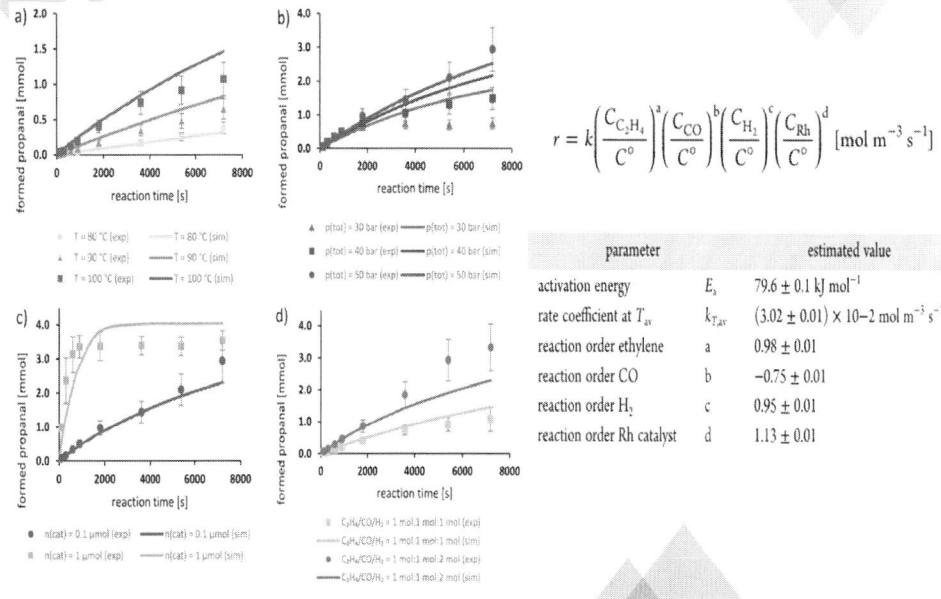

$$r = k\left(\frac{C_{C_2H_4}}{C^o}\right)^a \left(\frac{C_{CO}}{C^o}\right)^b \left(\frac{C_{H_2}}{C^o}\right)^c \left(\frac{C_{Rh}}{C^o}\right)^d \ [\text{mol m}^{-3}\text{ s}^{-1}]$$

parameter		estimated value
activation energy	E_a	79.6 ± 0.1 kJ mol^{-1}
rate coefficient at T_{av}	$k_{T,av}$	$(3.02 \pm 0.01) \times 10^{-2}$ mol m^{-3} s^{-1}
reaction order ethylene	a	0.98 ± 0.01
reaction order CO	b	-0.75 ± 0.01
reaction order H$_2$	c	0.95 ± 0.01
reaction order Rh catalyst	d	1.13 ± 0.01

Tubular Reactors (Lab 촉매 반응기)

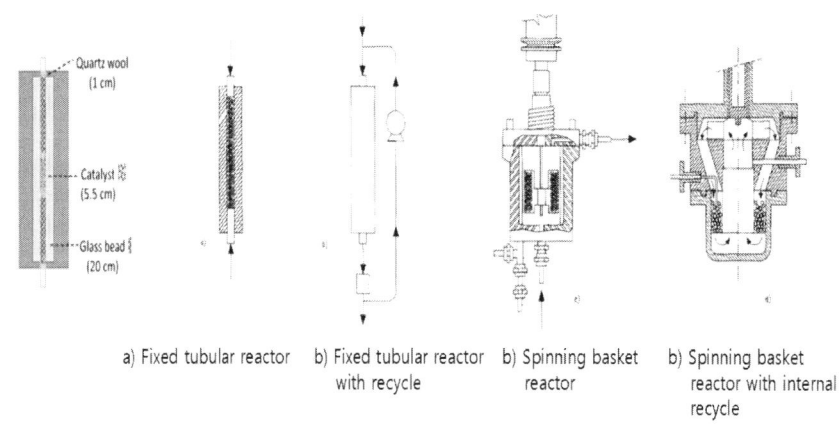

a) Fixed tubular reactor b) Fixed tubular reactor with recycle b) Spinning basket reactor b) Spinning basket reactor with internal recycle

Source : Catalysis Today 34 (1997)

Tubular Reactor Scale-up (촉매 반응)

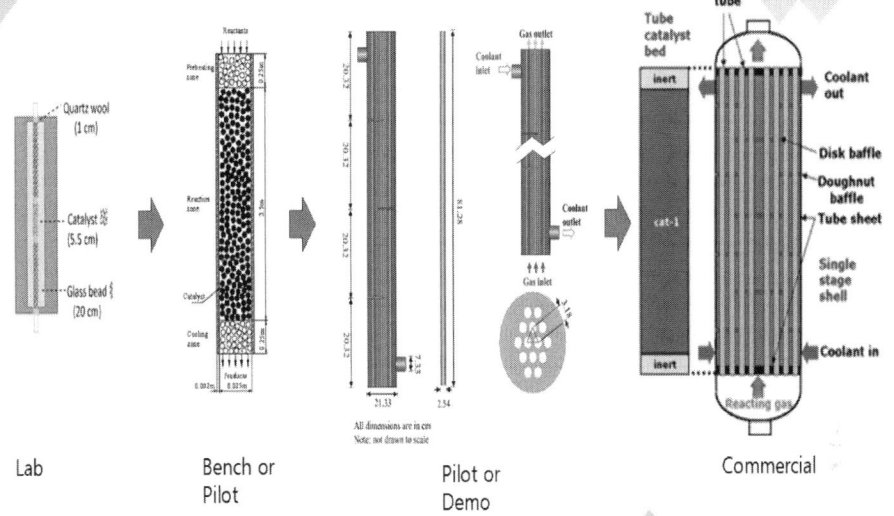

Lab   Bench or Pilot   Pilot or Demo   Commercial

Tubular Reactor Scale-up Issue (촉매 반응)

촉매의 다양한 형상   기공이 있는 구형 촉매에서의 Diffusion

Source : Catalyst Handbook   Source : Catalysis Today 34 (1997)

Kinetic model example (Partial oxidation)

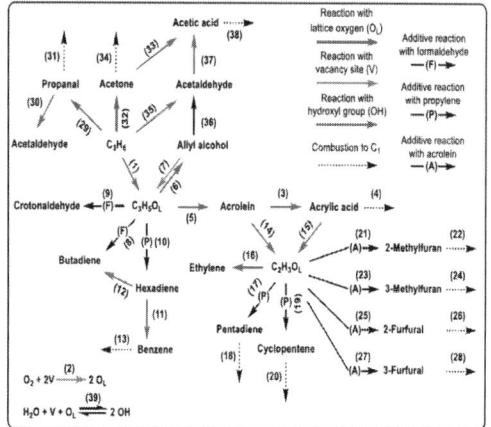

Kinetic model example (Partial oxidation)

$r_1 = 0.5 \times k_{C3H6-C3H5} P_{C_3H_6} \theta_{O_L}^2$

$r_2 = k_{O2} P_{O_2} \theta_V^2$

$r_3 = k_{C2H3CHO-C2H3COOH} P_{C_2H_3CHO} \theta_{O_L}^2$

$r_4 = k_{C2H3COOH-c} P_{C_2H_3COOH}$

$r_5 = 0.5 \times k_{C3H5-C2H3CHO} \theta_{C_3H_5O_L} \theta_{O_L}$

$r_6 = 0.5 \times k_{C3H5-C3H5OH} \theta_{C_3H_5} \theta_{OH}$

$r_7 = 0.5 \times k_{C3H5OH-C3H5} P_{C_3H_5OH} \theta_V \theta_{O_L}$

$r_8 = 0.5 \times k_{C3H5-C4H6} \theta_{C_3H_5} P_{HCHO}$

$r_9 = 0.5 \times k_{C3H5-C3H5CHO} \theta_{C_3H_5} P_{HCHO}$

$r_{10} = 0.5 \times k_{C3H5-C6H10} \theta_{C_3H_5} P_{C_3H_6}$

$r_{11} = k_{C6H10-C6H6} P_{C_6H_{10}} \theta_{O_L}^2$

$r_{12} = k_{C6H10-C4C2} P_{C_6H_{10}} \theta_{O_L}^2$

$r_{13} = k_{C6H6-c} P_{C_6H_6}$

$r_{14} = 0.5 \times k_{C2H3CHO-C2H3} P_{C_2H_3CHO} \theta_{O_L}^2$

$r_{15} = 0.5 \times k_{C2H3COOH-C2H3} P_{C_2H_3COOH} \theta_{O_L}^2$

$r_{16} = 0.5 \times k_{C2H3-C2H4} \theta_{C_2H_3} \theta_{OH}$

$r_{17} = 0.5 \times k_{C2H3-C5H8} \theta_{C_2H_3} P_{C_3H_6}$

$r_{18} = k_{C5H8-c} P_{C_5H_8}$

$r_{19} = 0.5 \times k_{C2H3-cC5H8} \theta_{C_2H_3} P_{C_3H_6}$

$r_{20} = k_{cC5H8-c} P_{cC_5H_8}$

$r_{21} = 0.5 \times k_{C2H3-2C4H3OCH3} \theta_{C_2H_3} P_{C_2H_3CHO}$

$r_{22} = k_{2C4H3POCH3-c} P_{2C_4H_3OCH_3}$

$r_{23} = 0.5 \times k_{C2H3-3C4H3OCH3} \theta_{C_2H_3} P_{C_2H_3CHO}$

$r_{24} = k_{3C4H3POCH3-c} P_{3C_4H_3OCH_3}$

$r_{25} = 0.5 \times k_{C2H3-3C4H3OCHO} \theta_{C_2H_3} P_{C_2H_3CHO}$

$r_{26} = k_{3C4H3POCHO-c} P_{3C_4H_3OCHO}$

$r_{27} = 0.5 \times k_{C2H3-2C4H3OCHO} \theta_{C_2H_3} P_{C_2H_3CHO}$

$r_{28} = k_{2C4H3POCHO-c} P_{2C_4H_3OCHO}$

$r_{29} = k_{C3H6-C2H5CHO} P_{C_3H_6} \theta_{OH}$

$r_{30} = k_{C2H5CHO-CH3CHO} P_{C_2H_5CHO} \theta_{O_L}^2$

$r_{31} = k_{C2H5CHO-c} P_{C_2H_5CHO}$

$r_{32} = k_{C3H6-CH3COCH3} P_{C_3H_6} \theta_{OH}$

$r_{33} = k_{CH3COCH3-CH3COOH} P_{CH_3COCH_3} \theta_{O_L}^2$

$r_{34} = k_{CH3COCH3-c} P_{CH_3COCH_3}$

$r_{35} = k_{C3H6-CH3CHO} P_{C_3H_4}$

$r_{36} = k_{C3H5OH-CH3CHO} P_{C_3H_5OH}$

$r_{37} = k_{CH3CHO-CH3COOH} P_{CH_3CHO} \theta_{O_L}^2$

$r_{38} = k_{CH3COOH-c} P_{CH_3COOH}$

Kinetic model example (Partial oxidation)

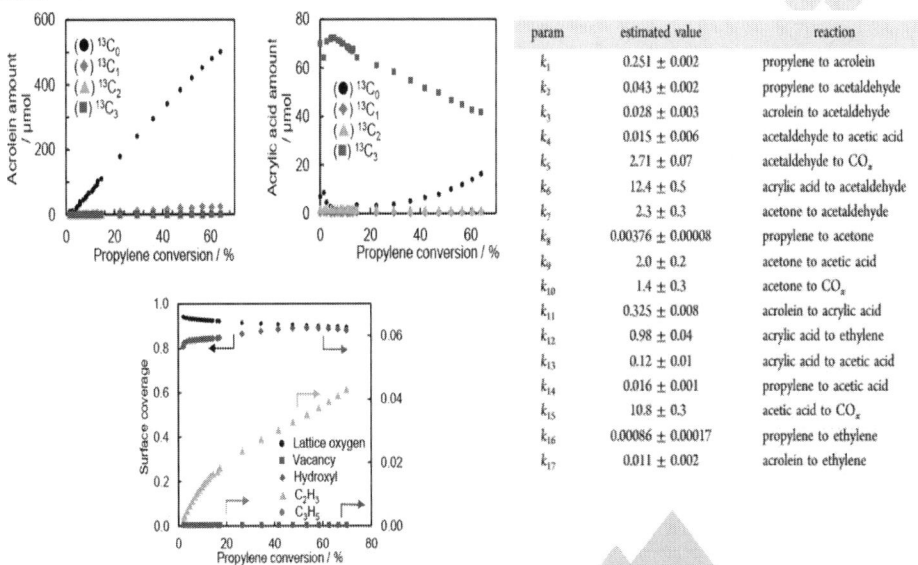

Packed Bed Reactor (Heat distribution)

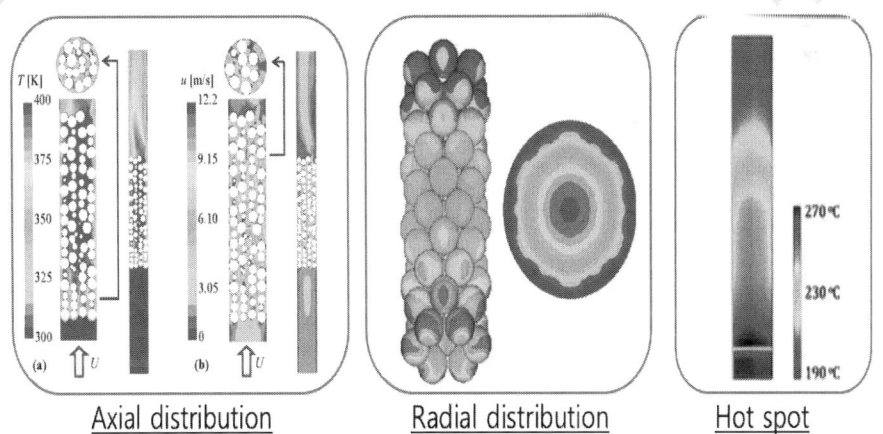

Axial distribution | Radial distribution | Hot spot

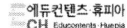

제11장
열교환기

Mechanism of Heat Transfer (1/2)

➢ **Conduction** $\dfrac{dQ}{dt} = -kA\dfrac{dT}{dx}$

where Q is the amount of heat, Btu, transferred in time t, h
k is the thermal conductivity, Btu/[h ft² (°F/ft)]
A is the area of heat transfer normal to heat flow, ft²
T is the temperature, °F
x is the thickness of the conduction path, ft.

➢ **Convection**

$$\dfrac{dQ}{dt} = hA\Delta T$$

h is the heat transfer coefficient, Btu/[h ft² °F]

Mechanism of Heat Transfer (2/2)

➢ **Radiation** $\dfrac{dQ}{dt} = \sigma \varepsilon AT^4$

where
σ is the Stefan-Boltzmann constant = 0.1713 10⁻⁸ Btu/(h ft² °R⁴)
ε is the emissivity of surface
A is the exposed area for heat transfer, ft²
T is absolute temperature, °R.

Rate of Heat Transfer

$$q = U A \Delta T_{tot}$$

U the overall heat transfer coefficient [=] Btu/(h ft² °F)
ΔT_{tot} is the total temperature difference (overall driving force for the process).

❖ The overall heat transfer coefficient, U, is an approximate value

❖ It is defined in combination with the area A
   (e.g. inside/outside area of pipe)

Rate of Heat Transfer

$$q = U A \Delta T_{tot}$$

✓ Assume :
   - no change in phase for either fluid
   - constant heat capacity for either fluid
   - negligible external heat loss from heat exchanger

▼

$$\dot{q} = -(\dot{m} C_p \Delta T)_h = (\dot{m} C_p \Delta T)_c$$

where $\dot{m}$ is the mass rate of flow of either the hot or cold streams
C_p is the heat capacity of these two streams.

Log Mean Temperature Difference

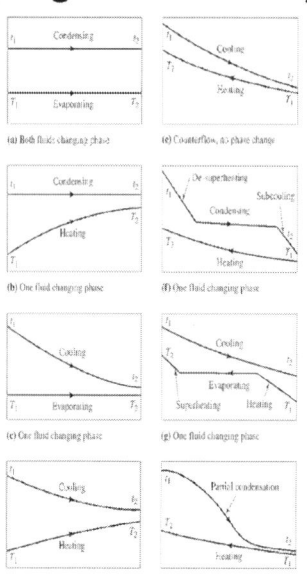

$$\Delta T_{o,\text{log mean}} = \frac{\Delta T_1 - \Delta T_2}{\ln(\Delta T_1/\Delta T_2)}$$

✓ When the temperature profiles are essentially linear

Overall Heat Transfer Coefficient

$$\frac{1}{U_o} = \frac{1}{h_o} + \frac{1}{h_{od}} + \frac{d_o \ln(d_o/d_i)}{2k_w} + \frac{d_o}{d_i}\frac{1}{h_i} + \frac{d_o}{d_i}\frac{1}{h_{id}}$$

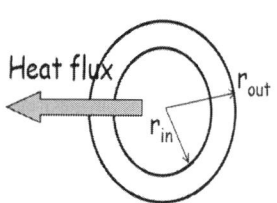

U_o: overall heat transfer coefficient based on the outside area
h_o, h_i: outside/inside film heat transfer coefficient
d_o, d_i: outside/inside pipe diameter
k_w: wall thermal conductivity
h_{od}, h_{id}: outside/inside fouling heat transfer coefficient
(=fouling factor)

What is Fouling?

> The most common definition of fouling in relation to heat exchangers is the deposition and accumulation of unwanted material such as scale, suspended solids, insoluble salts and even algae on the internal surfaces of the heat exchanger

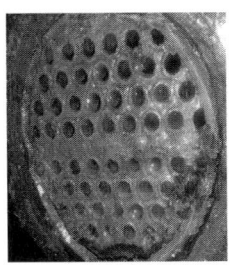  

The Fouling Effects

> **Heat-exchanger surfaces**

- Reduces thermal efficiency
- Increases T on the hot side, decreases T on the cold side
- Increases use of cooling water or cooling medium

> **Piping, flow channels**

- Reduces flow rate
- Increases pressure drop, and energy expenditure
- May cause flow oscillations, cavitation, and vibrations

MAINTENANCE IS NOT AN OPTION IT IS MUST!!!

참고) The Economic Impact of Fouling

- ✓ 8% of industry maintenance budget

- ✓ 0.25% of Western countries GDP, 0.20% of Chinese

(GDP, '21년)
- ❖ 미국 22.3조 USD
- ❖ 중국 17.7조 USD
- ❖ 일본 4.9조 USD
- ❖ 한국 1.8조 USD

- ✓ 20% increase in fuel cost for fossil fuel powered system

Source : www.watco-group.co/fouling-impact-economical/

Film Heat Transfer Coefficient

Fluids flowing inside of pipe (no phase change)

For viscous flow ($D_i G_i/\mu < 2100$),

$$h_i = 1.86 \frac{k}{D_i} \left(\frac{D_i G_i}{\mu}\right)^{1/3} \left(\frac{C_p \mu}{k}\right)^{1/3} \left(\frac{D_i}{L}\right)^{1/3} \left(\frac{\mu}{\mu_w}\right)^{0.14}$$

$$= 1.86 \left(\frac{4\dot{m}_i C_p}{\pi k L}\right)^{1/3} \left(\frac{\mu}{\mu_w}\right)^{0.14}$$

For turbulent flow above the transition region ($D_i G_i/\mu > 10,000, 0.7 < \text{Pr} < 160$ and $L/D_i > 10$)

$$h_i = 0.23 \frac{k}{D_i} \left(\frac{D_i G_i}{\mu}\right)^{0.8} \left(\frac{C_p \mu}{k}\right)^{1/3} \left(\frac{\mu}{\mu_w}\right)^{0.14}$$

In upper equations D_i is the inside diameter of the tube, G_i the mass velocity inside the tube, C_p the heat capacity of the fluid at constant pressure, k the thermal conductivity of the fluid, μ the viscosity of the fluid (subscript w indicates evaluation at the wall temperature), L the heated length of the straight tube, and m the mass rate of flow per tube. Physical properties k, C_p, and μ are evaluated at the average bulk temperature of the fluid.

Film Heat Transfer Coefficient

➢ Fluids flowing inside of pipe (no phase change)

For the transition region ($2100 < D_i G_i/\mu < 10{,}000$)

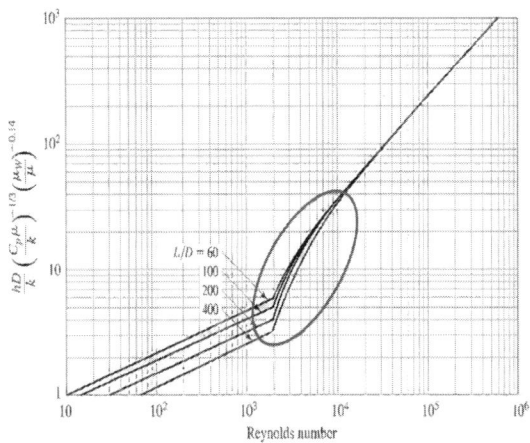

참고) Dimensionless number

➢ Reynolds No.

- It is the ratio of inertia force to the viscous force.

$$Re = \frac{\rho V L}{\mu}$$

Where,
- ρ is density
- V is velocity
- L is linear dimension
- μ is viscosity

Significance-It is used to identify the nature of flow (Laminar or Turbulent)

➢ Prandtl No.

- It is the ratio of momentum diffusivity to heat diffusivity
- Where

$$Pr = \frac{\mu C_p}{K}$$

- μ is the dynamic viscosity
- K is the thermal conductivity
- C_p is the specific heat

Significance-Prandtl number is used to describe thermal boundary layer.

➢ Nusselt No.

It is the ratio of convective heat transfer to conductive heat transfer.

Where
$$Nu = \frac{hL}{K}$$

- h is the convective heat transfer coefficient
- L is the characteristics length
- K is the thermal conductivity

Significance-It describe the enhancement of heat transfer because of convection in comparison to conduction

Film Heat Transfer Coefficient

➢ Fluids flowing outside of pipe (no phase change)

$$\frac{h_o D_o}{k} = 0.3 + \frac{0.62\,\text{Re}^{1/2}\text{Pr}^{1/3}}{[1+(0.4/\text{Pr})^{2/3}]^{1/4}}\left[1+\left(\frac{\text{Re}}{28,200}\right)^{5/8}\right]^{4/5}$$

General range of coefficients

General range of heat-transfer coefficients and fouling resistances for sensible heat transfer in tubular exchangers[a]

Fluid conditions	h, W/m²·K[b]	h_d^{-1}, m²·K/W
Water, liquid	$5\times10^3 - 1\times10^4$	$1\times10^{-4} - 2.5\times10^{-4}$
Light organics,[c] liquid	$1.5\times10^3 - 2\times10^3$	$1\times10^{-4} - 2\times10^{-4}$
Medium organics,[d] liquid	$7.5\times10^2 - 1.5\times10^3$	$1.5\times10^{-4} - 4\times10^{-4}$
Heavy organics,[e] liquid, heating	$2.5\times10^2 - 7.5\times10^2$	$2\times10^{-4} - 1\times10^{-3}$
Heavy organics,[e] liquid, cooling	$1.5\times10^2 - 4\times10^2$	$2\times10^{-4} - 1\times10^{-3}$
Very heavy organics,[f] liquid, heating	$1\times10^2 - 3\times10^{2\,g}$	$4\times10^{-4} - 3\times10^{-3}$
Very heavy organics,[f] liquid, cooling	$6\times10 - 1.5\times10^{2\,g}$	$4\times10^{-4} - 3\times10^{-3}$
Gas,[h] $p = 100$–200 kPa	$8\times10 - 1.2\times10^2$	$0 - 1\times10^{-4}$
Gas,[h] $p = 1$ MPa	$2.5\times10^2 - 4\times10^2$	$0 - 1\times10^{-4}$
Gas,[h] $p = 10$ MPa	$5\times10^2 - 8\times10^2$	$0 - 1\times10^{-4}$

[a]Data adapted from *Heat Exchanger Design Handbook*, G. F. Hewitt, ed., Begell House, New York, 1998, Sec. 3.1.4.
[b]Coefficients based on clean surface area in contact with the fluid and an allowable pressure drop of about 50–100 kPa for the fluid.
[c]Hydrocarbons through C$_8$, gasoline, light alcohols, ketones, etc.; $\mu < 0.5 \times 10^{-3}$ Pa·s
[d]Absorber oil, hot gas oil, kerosene, and light crudes; $0.5 \times 10^{-3} < \mu < 2.5 \times 10^{-3}$ Pa·s
[e]Lube oils, fuel oils, cold gas oil, heavy and reduced crudes; $2.5 \times 10^{-3} < \mu < 5.0 \times 10^{-2}$ Pa·s
[f]Tars, asphalts, greases, polymer melts, etc.; $\mu > 5.0 \times 10^{-2}$ Pa·s
[g]Estimation of coefficients is approximate and depends strongly on the temperature difference.
[h]Air, nitrogen, oxygen, carbon dioxide, light hydrocarbon mixtures, etc.

General range of coefficients (1/2)

General range of heat-transfer coefficients and fouling resistances for condensation and vaporization processes[a]

Fluid conditions	h, W/m²·K[b]	h_d^{-1}, m²·K/W
Steam, dropwise condensation	$5 \times 10^4 - 1 \times 10^5$	$0 - 1 \times 10^{-4}$
Steam, condensation, $p = 10$ kPa	$8 \times 10^3 - 1.2 \times 10^4$	$0 - 1 \times 10^{-4}$
Steam, condensation, $p = 100$ kPa	$1 \times 10^4 - 1.5 \times 10^4$	$0 - 1 \times 10^{-4}$
Steam, condensation, $p = 1$ MPa	$1.5 \times 10^4 - 2.5 \times 10^4$	$0 - 1 \times 10^{-4}$
Light organics,[c] condensation, $p = 10$ kPa	$1.5 \times 10^3 - 2 \times 10^3$	$0 - 1 \times 10^{-4}$
Light organics,[c] condensation, $p = 100$ kP	$2 \times 10^3 - 4 \times 10^3$	$0 - 1 \times 10^{-4}$
Light organics,[c] condensation, $p = 1$ Mpa	$3 \times 10^3 - 7 \times 10^3$	$0 - 1 \times 10^{-4}$
Medium organics,[d] condensation, $p = 100$ kPa, pure or narrow condensing range	$1.5 \times 10^3 - 4 \times 10^3$	$1 \times 10^{-4} - 3 \times 10^{-4}$
Heavy organics,[e] narrow condensing range, $p = 100$ kPa	$6 \times 10^2 - 2 \times 10^3$	$2 \times 10^{-4} - 5 \times 10^{-4}$
Light organic mixtures,[c] medium condensing range, $p = 100$ kPa	$1 \times 10^3 - 2.5 \times 10^3$	$0 - 2 \times 10^{-4}$
Medium organic mixtures,[d] medium condensing range, $p = 100$ kPa	$6 \times 10^2 - 1.5 \times 10^3$	$1 \times 10^{-4} - 4 \times 10^{-4}$
Heavy organic mixtures,[e] medium condensing range, $p = 100$ kPa	$3 \times 10^2 - 6 \times 10^2$	$2 \times 10^{-4} - 8 \times 10^{-4}$

General range of coefficients (2/2)

(Continued)

Fluid conditions	h, W/m²·K[b]	h_d^{-1}, m²·K/W
Light organic mixtures,[c] vaporization, $p < 2$ MPa, ΔT superheat max. = 15°C, narrow boiling range	$7.5 \times 10^2 - 3 \times 10^3$	$1 \times 10^{-4} - 3 \times 10^{-4}$
Medium organics,[d] vaporization, $p < 2$ MPa, ΔT superheat max. = 20°C	$1 \times 10^3 - 3.5 \times 10^3$	$1 \times 10^{-4} - 3 \times 10^{-4}$
Medium organic mixtures,[d] vaporization, $p < 2$ MPa, ΔT superheat max. = 15°C, narrow boiling range	$6 \times 10^2 - 2.5 \times 10^3$	$1 \times 10^{-4} - 3 \times 10^{-4}$
Heavy organics,[e] vaporization, $p < 2$ MPa, ΔT superheat max. = 20°C	$7.5 \times 10^2 - 2.5 \times 10^3$	$2 \times 10^{-4} - 5 \times 10^{-4}$
Heavy organic mixtures,[e] vaporization, $p < 2$ MPa, ΔT superheat max. = 15°C, narrow boiling range	$4 \times 10^2 - 1.5 \times 10^3$	$2 \times 10^{-4} - 8 \times 10^{-4}$
Very heavy organic mixtures,[g] vaporization, $p < 2$ MPa, ΔT superheat max. = 15°C, narrow boiling range	$3 \times 10^2 - 1 \times 10^{3\,f}$	$2 \times 10^{-4} - 1 \times 10^{-3}$

[a]Data adapted from *Heat Exchanger Design Handbook*, G. F. Hewitt, ed., Begell House, New York, 1998, Sec. 3.1.4.
[b]Coefficients based on clean surface area in contact with the fluid and an allowable pressure drop of about 50–100 kPa for the fluid.
[c]Hydrocarbons through C_8, gasoline light alcohols, ketones, etc.; $\mu < 0.5 \times 10^{-3}$ Pa·s
[d]Absorber oil, hot gas oil, kerosene, and light crudes; $0.5 \times 10^{-3} < \mu < 2.5 \times 10^{-3}$ Pa·s
[e]Lube oils, fuel oils, cold gas oil, heavy and reduced crudes; $2.5 \times 10^{-3} < \mu < 5.0 \times 10^{-2}$ Pa·s
[f]Estimation of coefficients is approximate.
[g]Tars, asphalts, greases, polymer melts, etc.; $\mu > 5.0 \times 10^{-2}$ Pa·s

Approximate Design Values of "U"

Hot Bid	Cold Bid	U_d, W/m²·K	h_d^{-1}, m²·K/W
Coolers			
Water	Water	1250 – 2500	2×10^{-4}
Methanol	Water	1250 – 2500	2×10^{-4}
Ammonia	Water	1250 – 2500	2×10^{-4}
Aqueous solutions	Water	1250 – 2500	2×10^{-4}
Light organics[†]	Water	375 – 750	6×10^{-4}
Medium organics[‡]	Water	250 – 600	6×10^{-4}
Heavy organics[§]	Water	25 – 375	6×10^{-4}
Gases	Water	10 – 250	6×10^{-4}
Water	Brine	500 – 1000	6×10^{-4}
Light organics[†]	Brine	200 – 500	6×10^{-4}
Heaters			
Steam	Water	1000 – 3500	2×10^{-4}
Steam	Methanol	1000 – 3500	2×10^{-4}
Steam	Ammonia	1000 – 3500	2×10^{-4}
Steam	Aqueous solutions:		
	$\mu < 2 \times 10^{-3}$ Pa·s	1000 – 3500	2×10^{-4}
	$\mu > 2 \times 10^{-3}$ Pa·s	500 – 2500	2×10^{-4}
Steam	Light organics[†]	500 – 1000	6×10^{-4}
Steam	Medium organics[‡]	250 – 500	6×10^{-4}
Steam	Heavy organics[§]	30 – 300	6×10^{-4}
Steam	Gases	25 – 250	6×10^{-4}
Utility fluid (e.g., Dowtherm)	Gases	20 – 200	6×10^{-4}
Utility fluid (e.g., Dowtherm)	Heavy organics[§]	30 – 300	6×10^{-4}

Hot Bid	Cold Bid	U_d, W/m²·K	h_d^{-1}, m²·K/W
Exchangers (no phase change)			
Water	Water	1400 – 2850	2×10^{-4}
Aqueous solutions	Aqueous solutions	1400 – 2850	2×10^{-4}
Light organics[†]	Light organics[†]	300 – 425	6×10^{-4}
Medium organics[‡]	Medium organics[‡]	100 – 300	6×10^{-4}
Heavy organics[§]	Heavy organics[§]	50 – 200	6×10^{-4}
Heavy organics[§]	Heavy organics[§]	150 – 300	6×10^{-4}
Light organics[†]	Heavy organics[§]	50 – 200	6×10^{-4}

[†]Hydrocarbons through C_8, gasoline, light alcohols, ketones, etc.; $\mu < 0.5 \times 10^{-3}$ Pa·s
[‡]Absorber oil, hot gas oil, kerosene, and light crudes; $0.5 \times 10^{-3} < \mu < 2.5 \times 10^{-3}$ Pa·s
[§]Lube oils, fuel oils, cold gas oil, heavy and reduced crudes; $2.5 \times 10^{-3} < \mu < 5.0 \times 10^{-3}$ Pa·s

Another Consideration for H/E

➤ Pressure drop

✓ Friction

- Resulting from flow of fluids through the H/E tubes and shell
- Due to sudden expansion and contraction

  - ❖ Tube-side pressure drop
  - ❖ Shell-side pressure drop

✓ Changes in vertical head and kinetic energy

- These effects are ordinarily relatively small
- To be neglected on many design calculation

Typical configuration of shell & tube HE

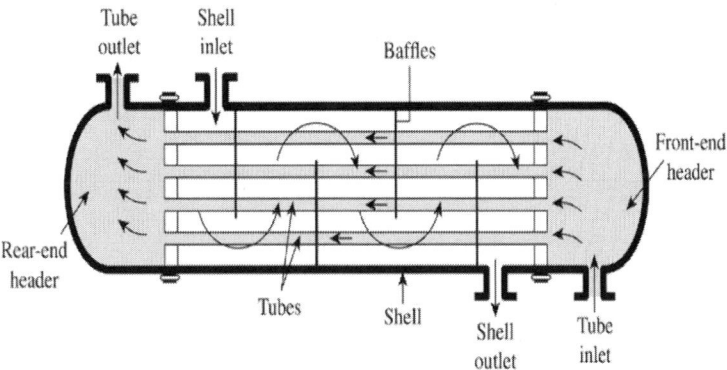

Various Types of shell & tube HE

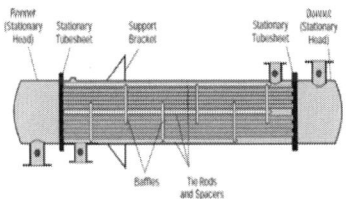

- Fixed tube type

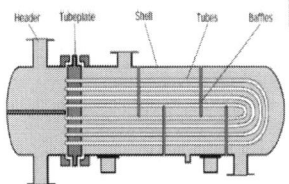

- U-tube type

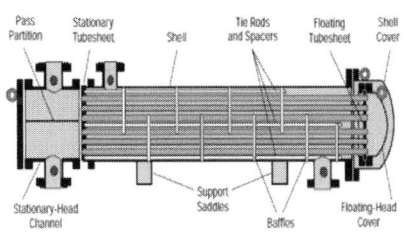

- Floating head type

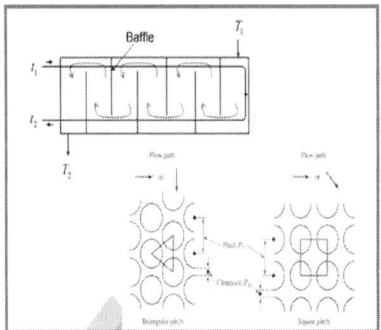

참고) TEMA

✓ **TEMA :**
(Tubular Exchanger
Manufactures Association)
- 원통다관형 열교환기 제조자 협회
- 열교환기 표준을 정립, 발행 및 Software도 배포하고 있다.
- 이 협회에서 발행하는 표준집을 TEMA Standard로 부름
- 열교환기 분야의 중요 Reference로 활용

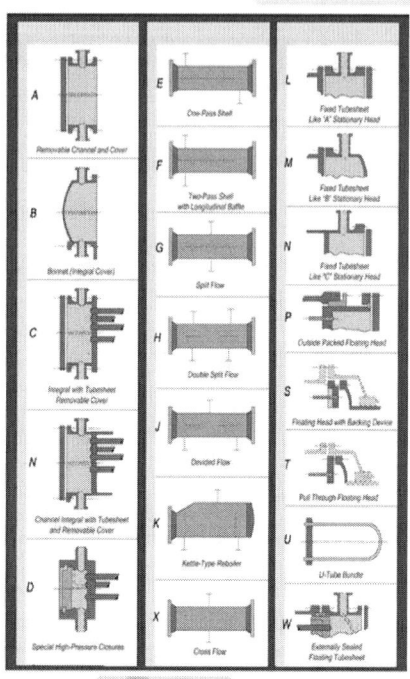

Various Types of HE (Plate type)

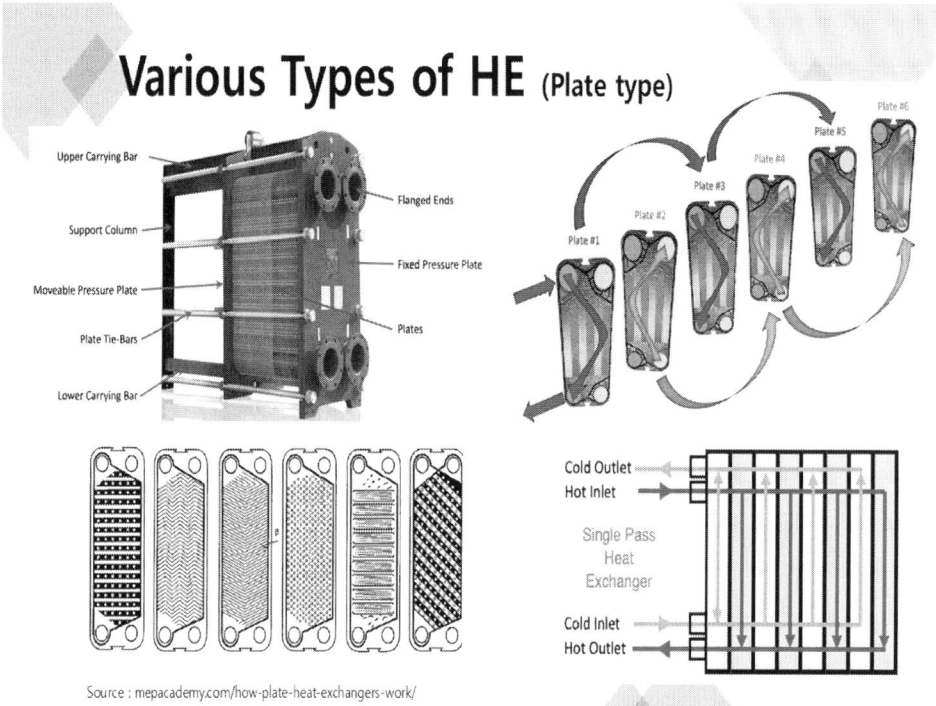

Source : mepacademy.com/how-plate-heat-exchangers-work/

Various Types of HE (Spiral type)

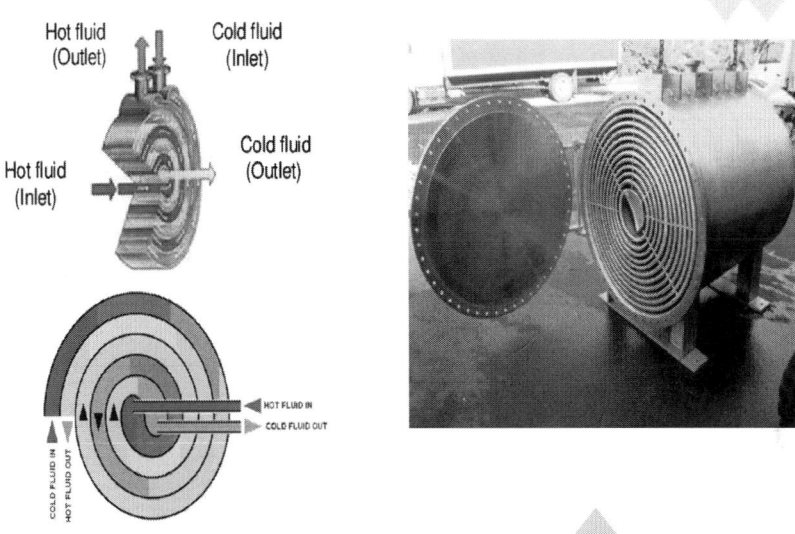

Source : www.elancoheatexchangers.com/products/spiral-plate-heat-exchangers/

The Process of HE design logic

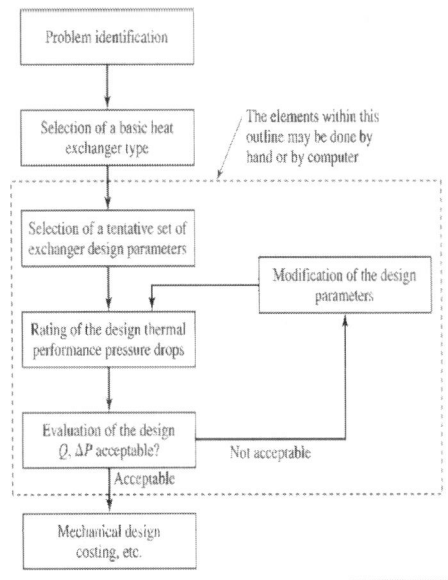

The Procedure of HE design

1) Define duty: heat transfer rate, flows, temperatures.
2) Collect required physical properties (ρ, μ, k).
3) Decide on the type of exchanger.
4) Select a trial value for U.
5) Calculate the mean temperature difference, ΔT_m
6) Calculate area required.
7) Decide on the exchanger layout.
8) Calculate individual coefficients.
9) Calculate U. If significant difference from step (4), substitute in (4) and repeat.
10) Calculate the pressure drop. If it is not satisfactory, back to (7) or (4) or (3).
11) Optimise: repeat (4) to (10) to determine cheapest solution (usually smaller area).

The General Considerations for HE

➢ Shell & Tube type (1/2)

Tube size:
Length is standard, commonly 8, 12 or 16 ft.
Diameter: most common 3/4 or 1 in OD

Tube pitch and clearance:
Pitch is the shortest center-to-center distance between adjacent tubes. Commonly 1.25 to 1.5 time the tube diameter.

Clearance is the distance between tubes. It should be larger than 25% of the tube diameter.

Triangular or square arrangement of tubes are quite common.

The General Considerations for HE

> Shell & Tube type (2/2)

Shell:
Up to 24 in nominal size, use standard pipes.

Baffles:
Baffles are usually spaced between 20% and 100% of the ID of the shell.

Fluid location:
Corrosive fluids flow inside the tubes.
Fluid with higher fouling tendency inside the tubes.
High pressure fluid inside the tubes (if everything else the same).
Hot fluid inside the tubes.

참고) Steel tube standard & Shell thickness

> Standard dimension for steel tubes

Outside Diameter (mm)	Wall Thickness (mm)				
16	1.2	1.7	2.1	—	—
19	—	1.7	2.1	2.8	—
25	—	1.7	2.1	2.8	3.4
32	—	1.7	2.1	2.8	3.4
38	—	—	2.1	2.8	3.4
50	—	—	2.1	2.8	3.4

> Minimum Shell thickness (mm)

Nominal Shell Dia., mm	Carbon Steel		Alloy Steel
	Pipe	Plate	
150	7.1	—	3.2
200–300	9.3	—	3.2
330–580	9.5	7.9	3.2
610–740	—	7.9	4.8
760–990	—	9.5	6.4
1010–1520	—	11.1	6.4
1550–2030	—	12.7	7.9
2050–2540	—	12.7	9.5

Heat exchanger Specification sheet

Cost comparison for HE type (1/2)

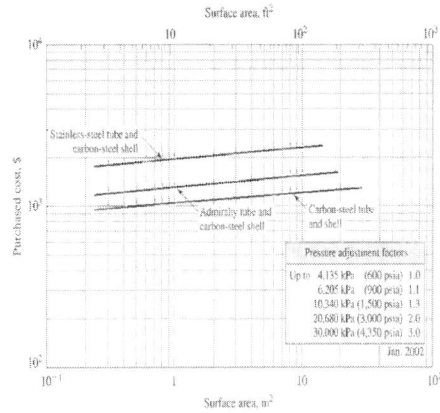

- Double-pipe HE

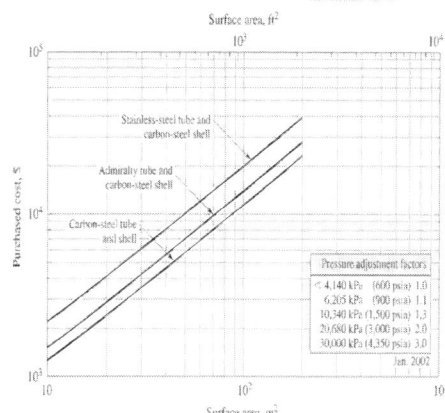

- Multiple-pipe HE

Cost comparison for HE type (2/2)

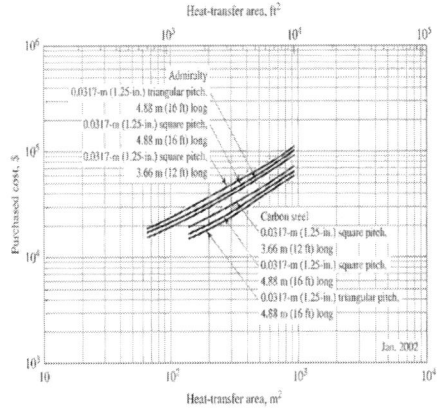

- Floating head HE

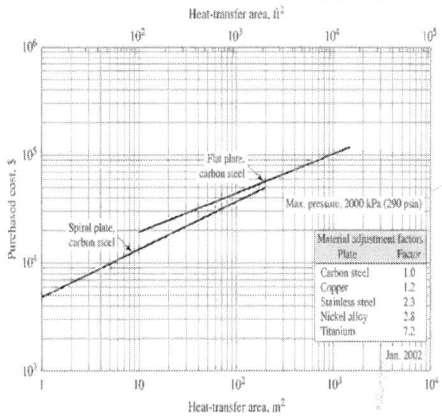

- Spiral & Plate HE

강 의 노 트

1.

2.

3.

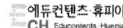

제12장

분리장치

분리장치

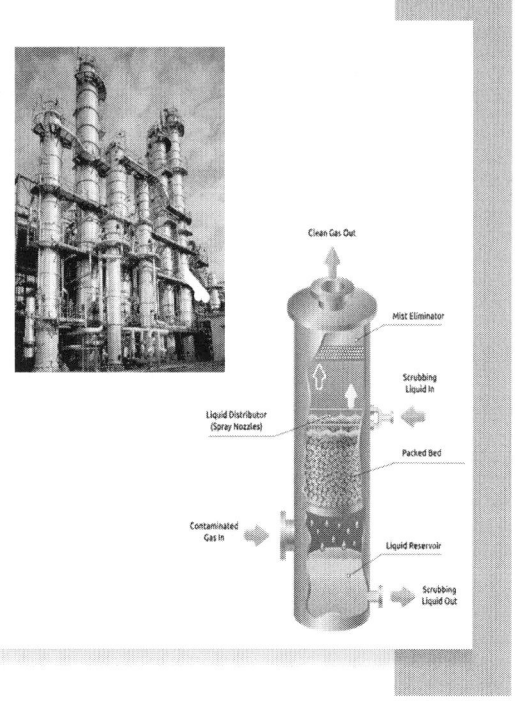

화학 반응

1) $A + B \rightarrow C$
2) $A + B \rightarrow C + D$ ⎫ Ideal Case

In Reality

➢ $A + B \rightarrow {}_A + {}_B + C + D + {}_E + {}_F + \cdots$

Considerations : G-L-S, T, P,
Physical properties etc.

화학공학의 독특한 2개 분야

> **Reaction : 반응공학**

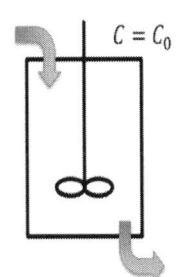

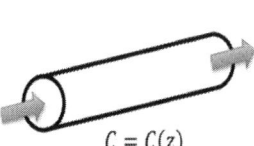

> **Separation : Transport Phenomena**

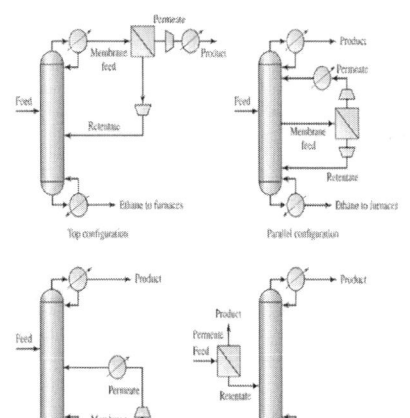

Various separation processes

- Distillation
- Absorption
- Stripping
- Extraction
- Crystallization
- Membrane separation

- Drying
- Evaporation
- Filtration
- Settling and Sedimentation
- Floatation
- Centrifugation

Properties associated with separation processes (1/2)

Exploitable property differences	Related separation processes
Vapor pressure	Distillation
	Absorption and stripping
	Drying
	Evaporation
Solubility	Crystallization
	Leaching
Distribution coefficient	Solvent extraction
	Adsorption
Exchange equilibrium	Ion exchange
	Chromatography
Surface activity	Froth flatation
Molecular geometry	Membrane
	Dialysis

Properties associated with separation processes (1/2)

	Exploitable property differences	Related separation processes
(Continued)	Molecular kinetic energy	Mass diffusion
		Thermal diffusion
	Electrical field	Precipitation
	Particle size	Filtration
		Screening
		Settling
		Sedimentation
	Particle size and density	Centrifugation
		Classification
		Thickening
		Decantation
		Scrubbing

‡Modified from data presented by D. R. Woods, *Process Design and Engineering Practice*, Prentice-Hall, Englewood Cliffs, NJ, 1995.

General Guidelines for separating homogeneous mixtures (1/2)

Separation processes for homogeneous mixtures	Feed condition and approx. mass fraction of key component	Separating agent, force fld, or gradient	Normal feed capacity limits set by equip., kg/s	Key process advantages	Key process limitations
Distillation	Liquid and/or vapor $10^{-3} - 0.95$	Heat transfer	$10^{-2} - 100$	Simple flwsheet, low capital investment, easily scalable, most widely used	Needs adequate relative volatility and thermal stability of components, low energy efficiency
Azeotropic distillation	Liquid and/or vapor (dictated by azeotropic conc.)	Liquid entrainer plus heat transfer	$10^{-4} - 60$	Breaks azeotrope, operates at normal temperature and pressure, easily scalable	System more complex, requires recovery of entrainer, increased capital investment
Extractive distillation	Liquid and/or vapor $10^{-3} - 0.95$	Liquid solvent plus heat transfer	$10^{-4} - 60$	Handles low-relative-volatility mixtures, operates at normal temperature and pressure, easily scalable	Requires inexpensive solvent with high recovery, system more complex, increased capital investment
Absorption	Vapor $10^{-3} - 0.95$	Liquid absorbent	$10^{-4} - 50$	Good for recovery of soluble gases, medium capital investment	Requires low-cost solvent and recovery system, cross-contamination
Stripping	Liquid $10^{-3} - 0.75$	Stripping vapor	$10^{-4} - 50$	Strips volatile component from liquid, simple operating conditions	Requires recovery of volatile component, involves cross-contamination
Extraction	Liquid	Liquid solvent	$10^{-4} - 50$	Great flxibility in selective solvents, easily scalable	Mutual solubilities of other components, involves cross-contamination, solvent loss
Crystallization	Liquid	Heat transfer	$10^{-4} - 10$	Single processing step at low temperatures, low energy requirement, large capacity variation	Staging not easy, may require parallel units, limited to crystal-forming components

General Guidelines for separating homogeneous mixtures (2/2)

(Continued)

Separation processes for homogeneous mixtures	Feed condition and approx. mass fraction of key component	Separating agent, force fld, or gradient	Normal feed capacity limits set by equip., kg/s	Key process advantages	Key process limitations
Membrane separation		Pressure gradient plus:		Good for bulk separation, clean air/water purification, some trace minerals, wide variety of membranes with range of selectivities	Low to moderate feed rates, requires chemically stable membrane, not easily staged, needs repressurization between stages, limited to non fouling flids
Microfiltration	Liquid	Microporous membrane	$10^{-5} - 1$		
Ultrafiltration	Liquid $3 \times 10^{-5} - 4 \times 10^{-3}$	Macroporous membrane			
Reverse osmosis	Liquid $7.5 \times 10^{-4} - 4 \times 10^{-2}$	Nonporous membrane	$10^{-5} - 5$		
Pervaporation	Liquid $5 \times 10^{-4} - 0.75$	Nonporous membrane	$10^{-5} - 1$		
Adsorption	Vapor or liquid $2 \times 10^{-3} - 2 \times 10^{-1}$	Solid adsorbent	$10^{-4} - 30$	High selectivity for low-concentration stream, good for gas purification	Requires recovery sequencing, needs high selectivity and easy regeneration
Electrical field and gradient separation	Vapor or liquid	Either centrifugal, electric force field or thermal gradient	Generally very low	Effective when electrical or gradient forces enhance the separation	Generally restricted to low flw rates, high equipment costs/unit capacity

[1] Modified from data presented by J. D. Seader and E. J. Henley, *Separation Process Principles*, J. Wiley, New York, 1988, G. D. Ulrich, *A Guide to Chemical Engineering Design and Economics*, J. Wiley, New York, 1984, J. C. Humphrey and G. E. Keller, II, *Separation Process Technology*, McGraw-Hill, New York, 1997, and D. R. Woods, *Process Design and Engineering Practice*, Prentice-Hall, Englewood Cliffs, NJ, 1995.

General Guidelines for separating heterogeneous mixtures (2/2)

Separation processes for heterogeneous mixtures	Feed condition and approx. mass fraction of key component	Separating agent, force fild, or gradient	Normal feed capacity limits set by equip., kg/s	Key process advantages	Key process limitations
Settling and sedimentation	Liquid or gas with solid $10^{-2} - 0.4$	Gravitational field	$10^{-3} - 2 \times 10^{1}$	Separation based on gravitational field, large capacities, low capital and maintenance costs	Poor filtrate clarity, high filtrate loss, requires large differences in density, poor with slime mixtures
Flotation	Liquid and solid $5 \times 10^{-4} - 4 \times 10^{-3}$	Adsorbent gas and surface activity	$10^{-3} - 4 \times 10^{-5}$	Useful in solid-solid separation, large capacity range	Separation based on selective gas bubble attachment, high operating cost to attain suitable particle size
Centrifugation	Solid-solid or liquid-solid $4 \times 10^{-1} - 0.3$	Centrifugal force field	$10^{-6} - 10^{-2}$	High separation efficiency	High capital and energy costs, limited capacity, discharge difficult
Drying	Wet solid $0.5 - 0.99$	Gas and/or heat transfer	$5 \times 10^{-1} - 2$	Final drying of solids once most, liquid has been removed	Staging inconvenient, may require parallel units, high energy requirement
Evaporation	Liquid	Heat transfer	$10^{-1} - 30$	Use when relative volatility >20	Low thermal efficiency
Condensation	Vapor	Heat transfer	$10^{-1} - 30$	Recovery of less volatile components in vapor or condensation of vapor	Appropriate coolant required
Filtration	Liquid-solid or gas-solid $10^{-2} - 0.75$	Pressure gradient plus porous filter	$10^{-6} - 10^{-4}$	Good solid-liquid separation of fibers, pulps, granules and slimes, low capital cost	High operating cost, often batch operation, poor with toxic or hazardous materials

*Modified from data presented by J. D. Seader and E. J. Henley, *Separation Process Principles*, J. Wiley, New York, 1988, G. D. Ulrich, *A Guide to Chemical Engineering Design and Economics*, J. Wiley, New York, 1984, J. C. Humphrey and G. E. Keller, II, *Separation Process Technology*, McGraw-Hill, New York, 1997, and D. R. Woods, *Process Design and Engineering Practice*, Prentice-Hall, Englewood Cliffs, NJ, 1995.

Distillation & Absorption tower

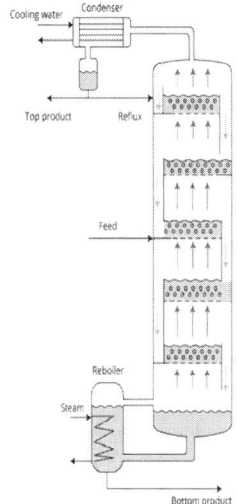

Source : chemengvirtual.uwaterloo.ca/distillation-lab/components/distillation-column.html

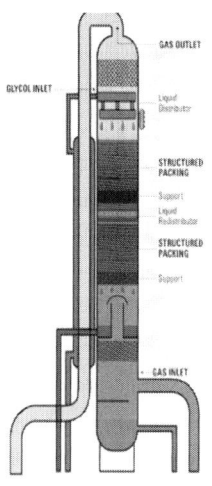

Source : kimray.com/training/gas-absorber-tower-how-3-types-work-dehydrate-your-natural-gas

Stripping & Extraction column

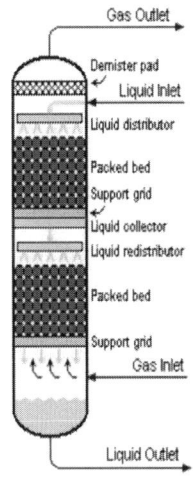

Source : shreewire.co.in/stripper-column/

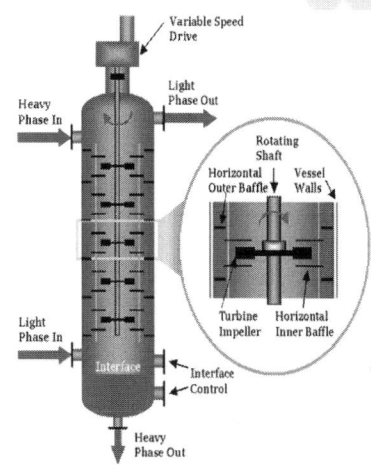

Source : kochmodular.com/liquid-liquid-extraction/extraction-column-types/

Dryer & Centrifuge

Source : www.okgemco.com/vacuum-dryers/double-cone

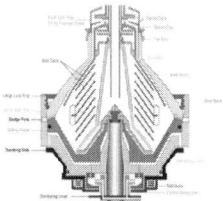

Source : dolphincentrifuge.com/industrial-centrifuge/

◈ 제12장. 분리장치 ▶ 277

Design procedure for Distillation column

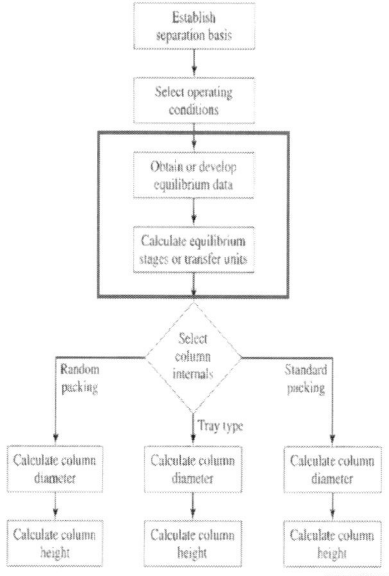

Basic Principles (Stage Equations)

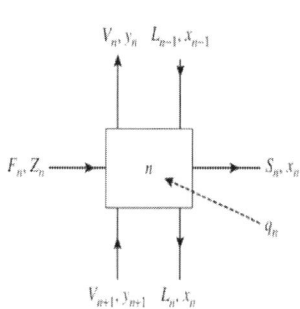

Material balance

$$V_{n+1}y_{n+1} + L_{n-1}x_{n-1} + F_n z_n = V_n y_n + L_n x_n + S_n x_n$$

Energy balance

$$V_{n+1}H_{n+1} + L_{n-1}h_{n-1} + Fh_f + q_n = V_n H_n + L_n h_n + S_n h_n$$

where V_n = vapor flow from the stage
V_{n+1} = vapor flow into the stage from the stage below
L_n = liquid flow from the stage
L_{n-1} = liquid flow into the stage from the stage above
F_n = any feed flow into the stage
S_n = any side stream from the stage
q_n = heat flow into, or removal from, the stage
n = any stage, numbered from the top of the column

From Equilibrium data to Column

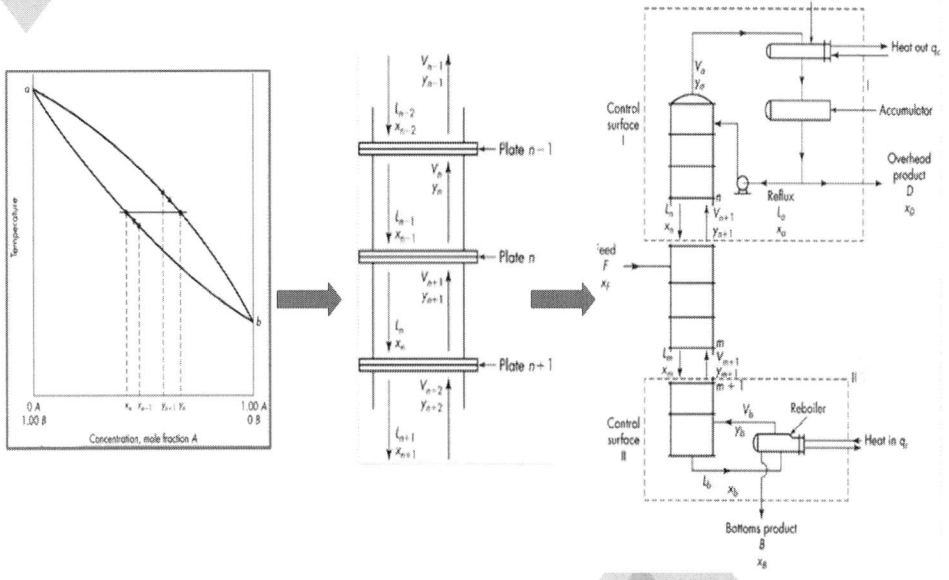

Number of Equilibrium Stage (Ideal)

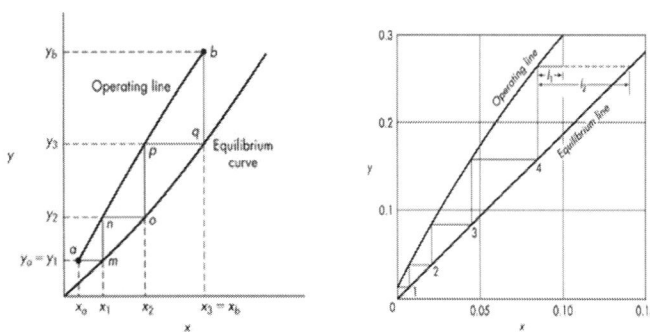

It is essential to get the fowling data for design of distillation
- *Vapor-Liquid Equilibrium data (V-L-E) or*
- *Binary Equilibrium data*

  ❖ *NTRL, UNIQUAC, UNIFAC model*

Various Equilibrium curves

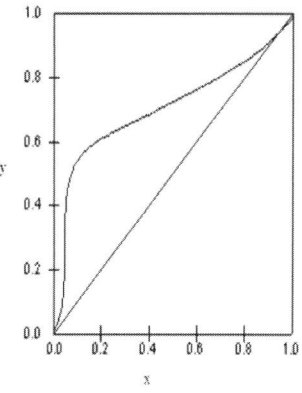

➢ Non-ideal systems

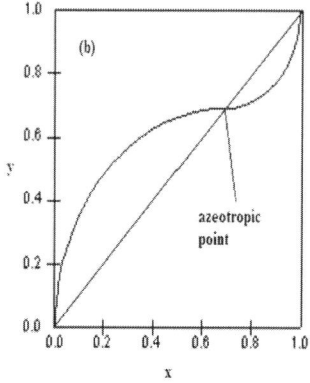

➢ Azeotropic systems

Configuration of Distillation column

- ➢ Overhead(Top) product
- ➢ Condenser
- ➢ Reflux/Splitter

- ➢ Rectifying section
- ➢ Feed stage
- ➢ Stripping section

- ➢ Bottom product
- ➢ Reboiler

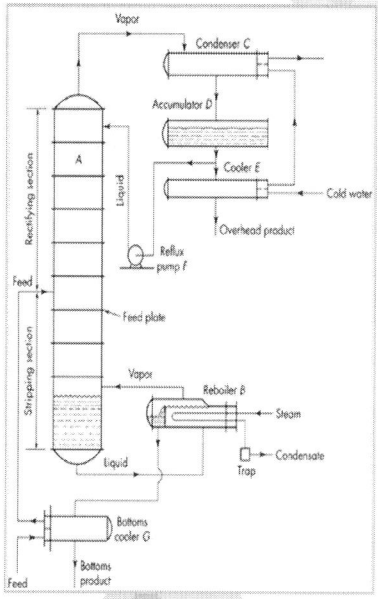

Distillation control (in general case)

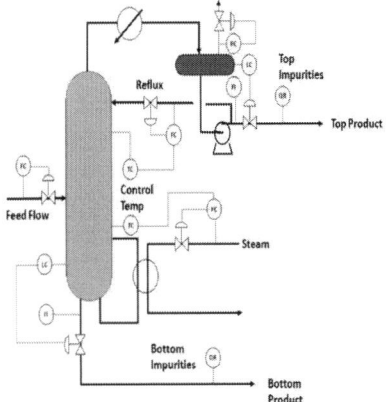

Control variables
> Composition and Physical properties of feed and product
> Flow rate, max and min T, P, ΔP

Various concept for Distillation

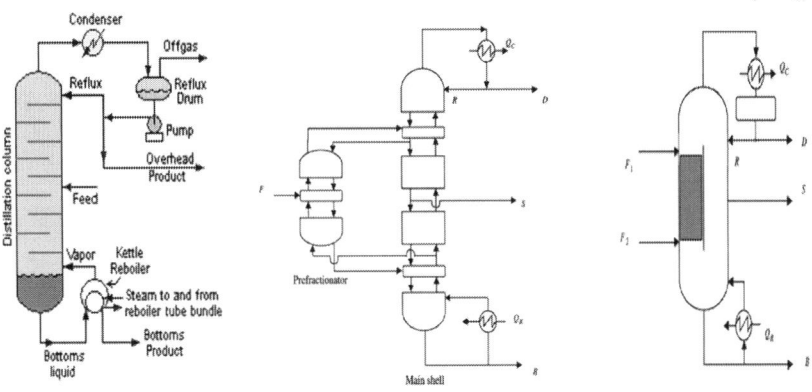

✓ Typical type ✓ Petlyuk column ✓ Divided wall column

Column sequences (simple)

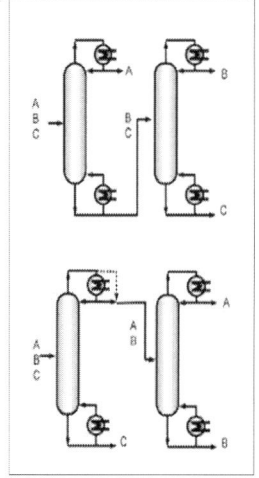

✓ For 3 components

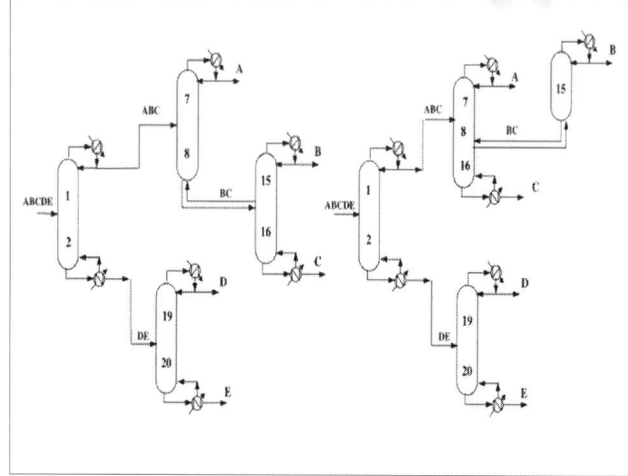

✓ For 5 components

Column sequences (complex)

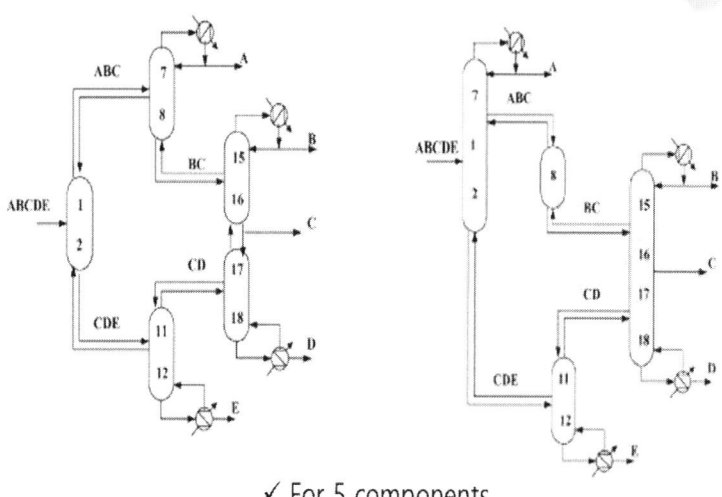

✓ For 5 components

Azeotropic & Extractive distillation

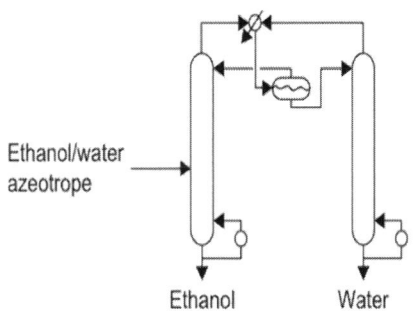

✓ Dehydration of Ethanol
 (using benzene as entrainer)

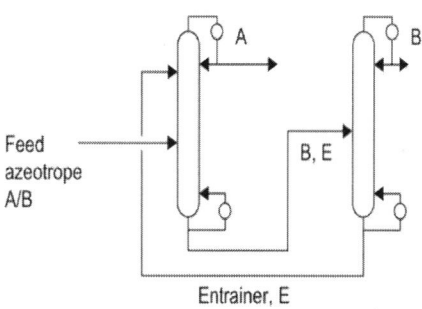

✓ Extractive distillation

Cross-sectional view in tray column

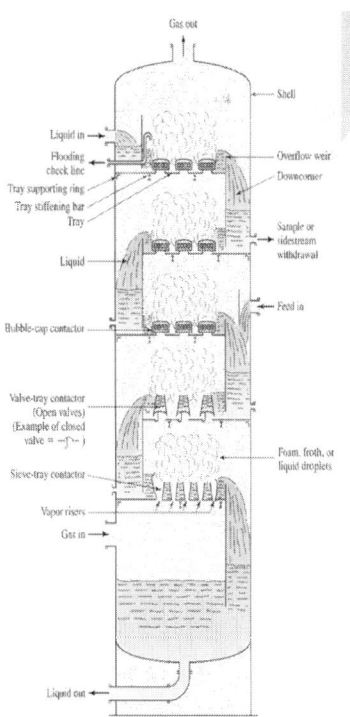

Typical inner view of column

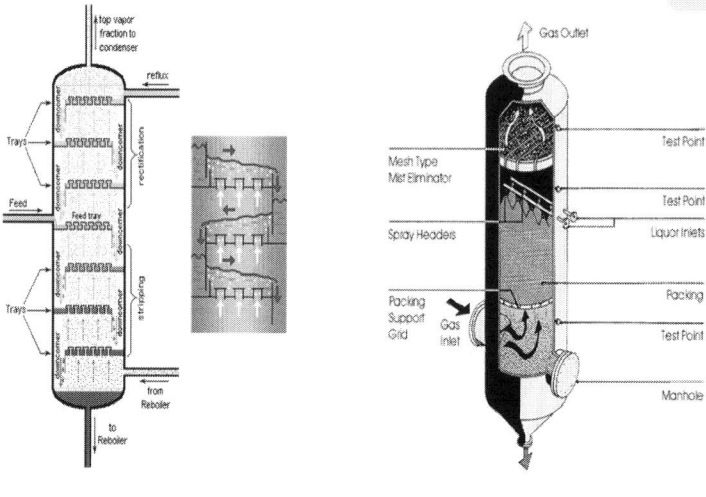

> Tray column

> Packed column

Typical sectional-plate

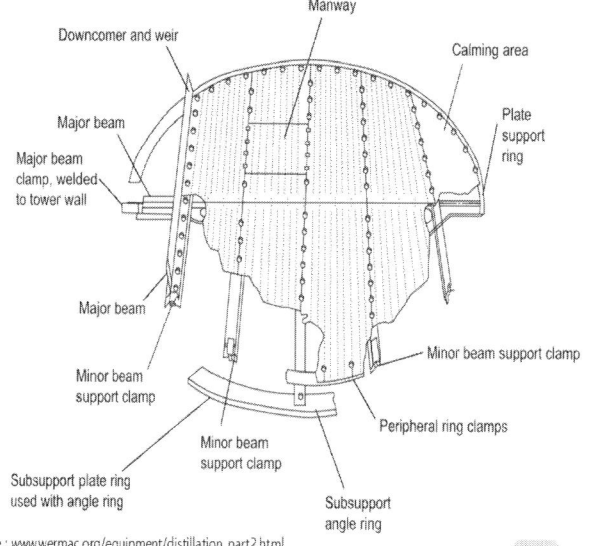

Source : www.wermac.org/equipment/distillation_part2.html

Various types of tray (1/3)

➢ Sieve tray

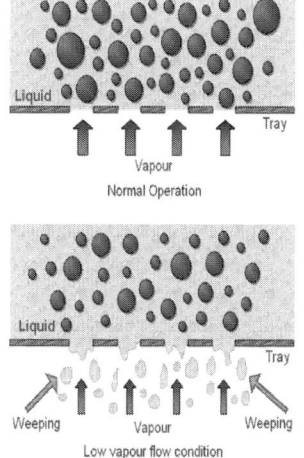

Source : www.wermac.org/equipment/distillation_part2.html

Various types of tray (2/3)

➢ Valve tray

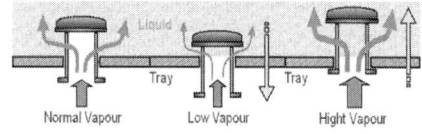

Source : www.wermac.org/equipment/distillation_part2.html

◆ 제12장. 분리장치 ▶ 285

Various types of tray (3/3)

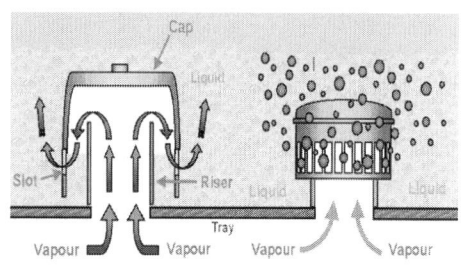

➢ Bubble cap

Source : www.wermac.org/equipment/distillation_part2.html

Operation Limits for Trays

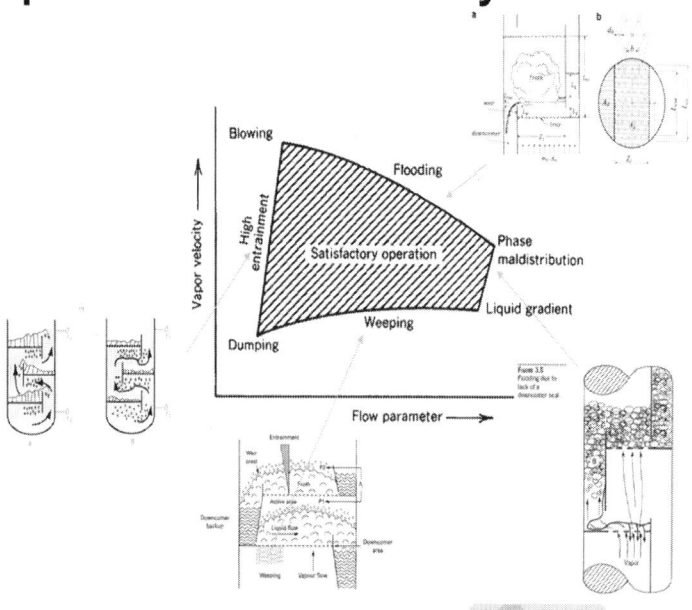

Design point for Tray

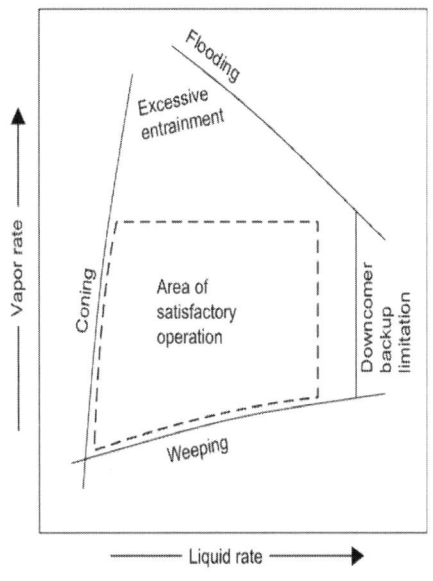

Design variables and procedure for tray column

- Diameter (w/ flooding velocity)
- Tray spacing : 12 or 24 inch
- Downcomer area : Minimum width 5 inch
- Hole diameter : 3/16 ~ 1/4 inch
- Total hole area $\frac{A_{All-holes}}{A_{holes}} = K\left(\frac{hole-diameter}{hole-pitch}\right)^2$ $K = 0.905$ (equilateral triangular pitch) $K = 0.785$ (rectangular pitch)
- Number of holes
- Height of weirs
- Pressure drop
- Efficiency

Overview of structured packing column

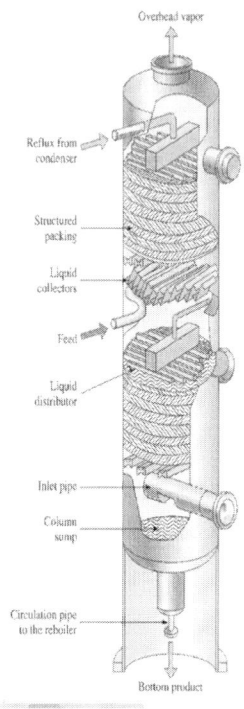

Various types of packing materials

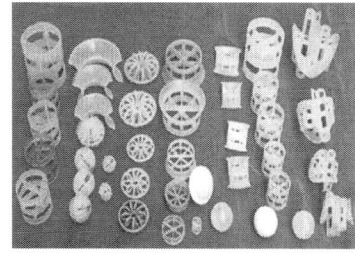

> Structured packing
> Fexipack

Source : amacs.com/blog/what-is-structured-packing-and-how-is-it-used-in-process-plants/

Design variables for packed column

> Packing Height : Number of equilibrium stages X HETP
  * HETP : Height Equivalent to a Theoretical Plate

> HETP : Typical a function of gas rate and the packing, as well as the mixture

> Packing Diameter

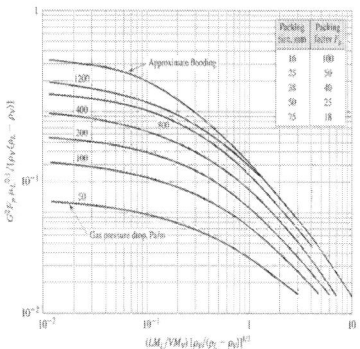

Maintenance for Distillation column

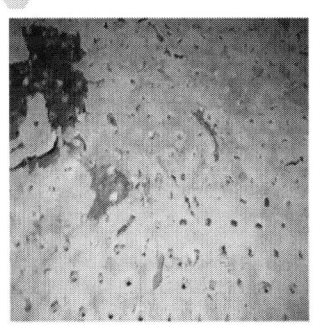
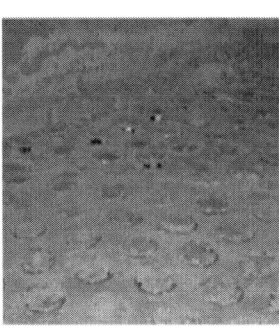

> Polymer in sieve tray   > Polymer in valve tray   > Scales at wall

Purchased Cost of Distillation column

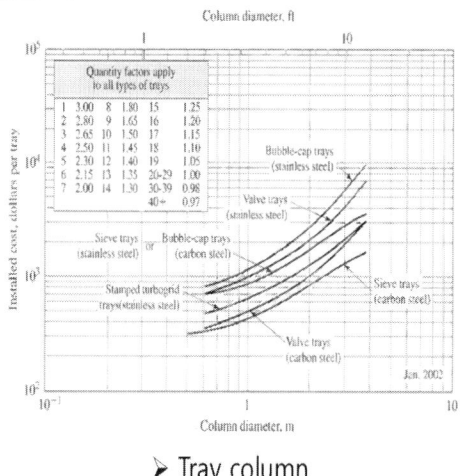

➢ Tray column

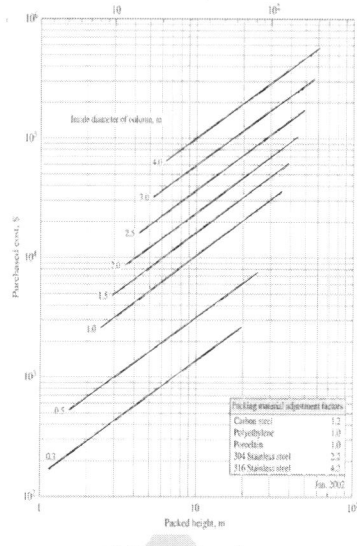

➢ Packed column

Purchased Cost of Distillation column
(계장 포함)

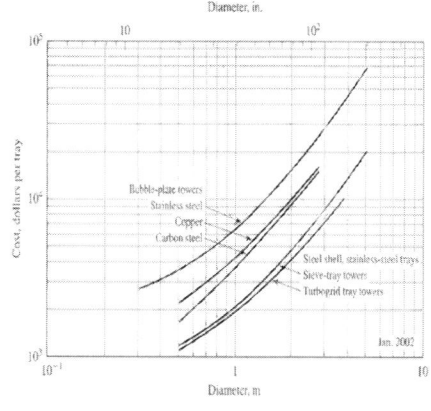

➢ Tray column

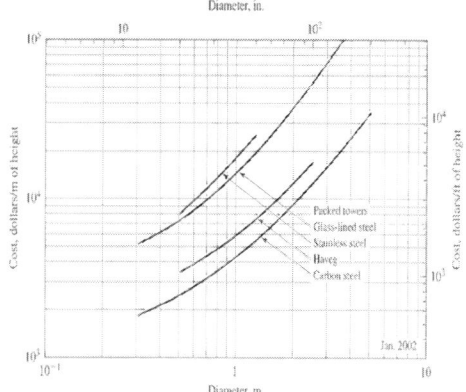

➢ Packed column

제13장
투자비 및 생산비용/매출

투자비 추정

Cash-flow diagram (1/2)

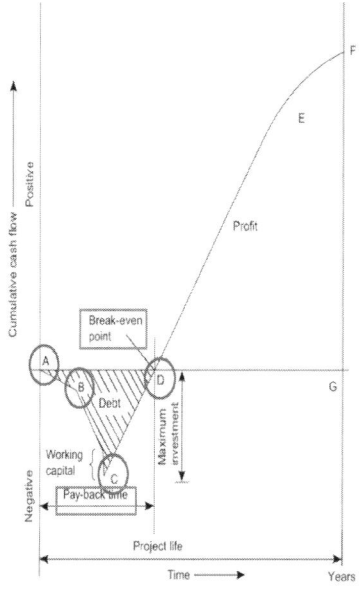

- A-B : The investment to the plant

- B-C : The heavy flow of capital to build the plant and provide funds for working capital

- C-D : The cash-flow curve turns up at C, as the process comes on stream and income is generated from sale

- The cumulative amount remains negative until the investment is paid at D. Point D is known as the ***break-even point*** and the time to reach the break-even point is called the ***payback time***

Cash-flow diagram (2/2)

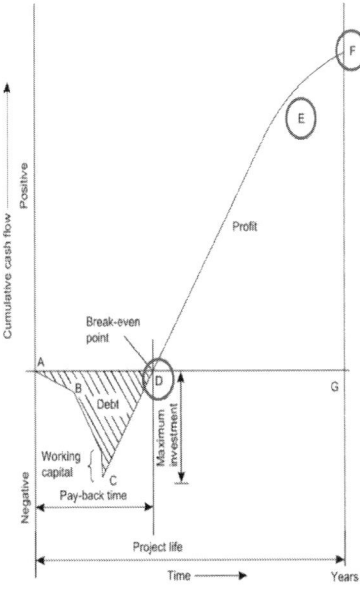

- D-E : The project is earning a return on the investment

- E-F : Toward the end of project life the rate of cash-flow may tend to fall off, due to increased operating costs and falling sales volume and price due to obsolescence of the plant

- The point F gives final cumulative net cash-flow at the end of the project life

What is Cost of Capital?

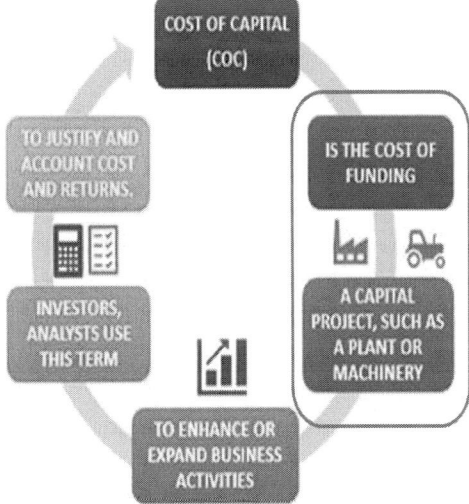

Source : www.wallstreetmojo.com/cost-of-capital/

Factors affecting Investment costs

- Cost of Equipment

- Price fluctuations (Economic market)

- Company policies (ex. : Safety, Accounting, Labor union, ...)

- Operating time & Rate of production
  (ex. : Maintenance, Depreciation, Sales demand, ...)

- Government policies
  (ex. : Import & Export tariff, Income tax, Environmental regulations, ...)

Ex.) The effect of costs and profit
(based on the rate of production)

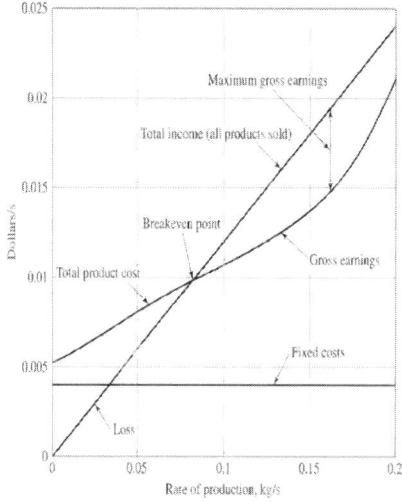

Break-even chart for some chemical plant

Capital Cost Investment

✓ Economic definition of **_capital_** is "stock of accumulated wealth"

✓ **_Investment_** is to promote the production of other goods
... based on the view of obtaining _income_ or _profit_

❖ Fixed Capital Investment (FCI, 고정자본 투자비)
  - to supply the required manufacturing and plant facilities

❖ Working Capital (WC, 운전자본)
  - necessary cost for the operation of the plant

◆ Total Investment Cost (TCI, 총 자본 투자비) = FCI + WC

Components of Capital cost

❖ Fixed Capital

The total cost of designing, constructing, and installing a plant and the associated modification needed to prepare the plant site

- ISBL
- OSBL
- Engineering and construction cost
- Contingency charges

* ISBL : Inside battery limit
* OSBL : Offsite battery limit

❖ Working Capital

The additional money to start the plant up and keep it running. It is the money that is tied up in sustaining plant operation

- Value of raw material inventory
- Value of product inventory
- Cash for cost of production (1wk)
- Accounts receivable – products shipped but not paid (1month)
- Credit for accounts payable – feedstocks, solvents etc.
- Spare parts inventory
  -1~2% ISBL+ OSBL

참고) ISBL and OSBL

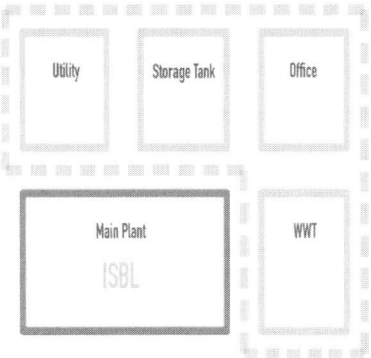
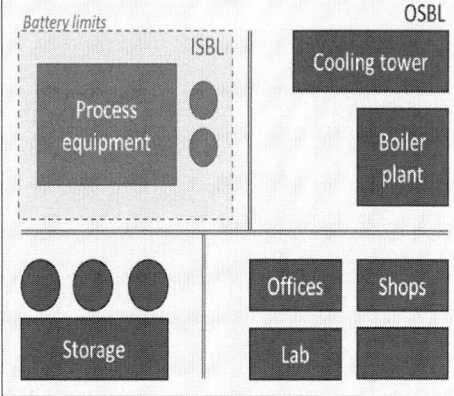

Fixed Capital : ISBL Costs

The costs of procuring and installing all the process equipment that makes up the new plant

The direct field costs include:

1. All the major process equipment, such as vessels, reactors, columns, furnaces, heat exchangers, coolers, pumps, compressors, motors, fans, turbines, filters, centrifuges, driers, etc., including field fabrication and testing if necessary
2. Bulk items, such as piping, valves, wiring, instruments, structures, insulation, paint, lube oils, solvents, catalysts, etc.
3. Civil works such as roads, foundations, piling, buildings, sewers, ditches, bunds, etc.
4. Installation labor and supervision

In addition to the direct field costs there will be indirect field costs including:

1. Construction costs such as construction equipment rental, temporary construction (rigging, trailers, etc.), temporary water and power, construction workshops, etc.
2. Field expenses and services such as field canteens, specialists' costs, overtime pay, and adverse weather costs
3. Construction insurance
4. Labor benefits and burdens (social security, workers compensation, etc.)
5. Miscellaneous overhead items such as agent's fees, legal costs, import duties, special freight costs, local taxes, patent fees or royalties, corporate overheads, etc.

Fixed Capital : OSBL Costs

The costs of the additions that must be made to the site infrastructure to accommodate adding a new plant or increasing the capacity of existing plant

- Electric main substations, transformers, switchgear, power lines
- Power generation plants, turbine engines, standby generators
- Boilers, steam mains, condensate lines, boiler feed water treatment plant, supply pumps
- Cooling towers, circulation pumps, cooling water mains, cooling water treatment
- Water pipes, water demineralization, waste water treatment plant, site drainage and sewers
- Air separation plants to provide site nitrogen for inert gas, nitrogen lines
- Dryers and blowers for instrument air, instrument air lines
- Pipe bridges, feed and product pipelines
- Tanker farms, loading facilities, silos, conveyors, docks, warehouses, railroads, lift trucks
- Laboratories, analytical equipment
- Offices, canteens, changing rooms, central control rooms
- Workshops and maintenance facilities
- Emergency services, firefighting equipment, fire hydrants, medical facilities, etc.
- Site security, fencing, gatehouses, landscaping

✓ For typical chemicals projects, OSBL : 20~50% of ISBL,
✓ 40% of ISBL usually used as an initial estimate if no details of the site are known

Fixed Capital : Engineering Costs

The costs of detailed design and other engineering services required to carry out the project

1. Detailed design engineering of process equipment, piping systems, control systems and offsites, plant layout, drafting, cost engineering, scale models, and civil engineering
2. Procurement of main plant items and bulks
3. Construction supervision and services
4. Administrative charges, including engineering supervision, project management, expediting, inspection, travel and living expenses, and home office overheads
5. Bonding
6. Contractor's profit

✓ A rule of thumb for engineering cost
  - for smaller project : 30% of ISBL+OSBL
  - for larger project : 10% of ISBL+OSBL

Fixed Capital : Contingency charges

Contingency charges are extra costs added into the project budget to allow for variation
From the cost estimate. Apart from in the cost estimate, contingency costs help cover

- Minor changes in project scope
- Changes in prices (e.g., prices of steel, copper, catalyst, etc.)
- Currency fluctuations
- Labor disputes
- Subcontractor problems
- Other unexpected problems

✓ A minimum contingency charge : 10% of ISBL+OSBL
  - If the technology is uncertain, contingency charges up to 50%

Working Capital

Working capital is the additional money needed, above what it cost to build the plant,
To start plant up and keep it running. Working capital is best thought of as the money
That is tied up in sustaining plant operation.

1. Value of raw material inventory—usually estimated as two weeks' delivered cost of raw materials
2. Value of product and by-product inventory—estimated as two weeks' cost of production
3. Cash on hand—estimated as one week's cost of production
4. Accounts receivable—products shipped but not yet paid for—estimated as one month's cost of production, but could be larger depending on customer payment terms
5. Credit for accounts payable—feedstocks, solvents, catalysts, packaging, etc., received but not yet paid for—estimated as one month's delivered cost, but could be larger depending on terms negotiated with vendors
6. Spare parts inventory—estimated as 1% to 2% of ISBL plus OSBL investment cost

✓ A typical figure for general chemicals and petrochemical plant : 15% of ISB+OSBL

Estimation of Capital Investment

1. *Order-of-magnitude estimate* (*ratio estimate*) based on similar previous cost data; probable accuracy of estimate over ±30 percent.
2. *Study estimate* (*factored estimate*) based on knowledge of major items of equipment; probable accuracy of estimate up to ±30 percent.
3. *Preliminary estimate* (*budget authorization estimate* or *scope estimate*) based on sufficient data to permit the estimate to be budgeted; probable accuracy of estimate within ±20 percent.
4. *Definitive estimate* (*project control estimate*) based on almost complete data but before completion of drawings and specifications; probable accuracy of estimate within ±10 percent.
5. *Detailed estimate* (*contractor's estimate*) based on complete engineering drawings, specifications, and site surveys; probable accuracy of estimate within ±5 percent.

The accuracy range for Cost-estimation

ESTIMATE CLASS	Primary Characteristic	Secondary Characteristic			
	MATURITY LEVEL OF PROJECT DEFINITION DELIVERABLES Expressed as % of complete definition	END USAGE Typical purpose of estimate	METHODOLOGY Typical estimating method	EXPECTED ACCURACY RANGE Typical +/- range relative to index of 1 (i.e. Class 1 estimate) [a]	PREPARATION EFFORT Typical degree of effort relative to least cost index of 1 [b]
Class 5	0% to 2%	Screening or feasibility	Stochastic (factors and/or models) or judgment	4 to 20	1
Class 4	1% to 15%	Concept study or feasibility	Primarily stochastic	3 to 12	2 to 4
Class 3	10% to 40%	Budget authorization or control	Mixed but primarily stochastic	2 to 6	3 to 10
Class 2	30% to 75%	Control or bid/tender	Primarily deterministic	1 to 3	5 to 20
Class 1	65% to 100%	Check estimate or bid/tender	Deterministic	1	10 to 100

Notes: [a] If the range index value of "1" represents +10/-5%, then an index value of 10 represents +100/-50%.
[b] If the cost index value of "1" represents 0.005% of project costs, then an index value of 100 represents 0.5%.

The typical series of Cost estimation

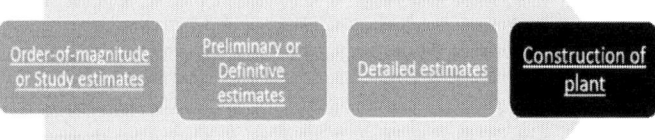

Order-of-magnitude or Study estimates → Preliminary or Definitive estimates → Detailed estimates → Construction of plant

❖ Basically, cost estimation is devoted to preliminary estimate

Cost-estimating information guide (1/3)

Required information		Detailed estimate ± 5% range	Definitive estimate ± 10% range	Preliminary estimate ± 20% range	Study estimate ± 30% range	Order-of-magnitude estimate > ± 30% range
Site	Location	●	●	●	●	
	General description	●	●	●	●	
	Soil bearing	●	●	●		
	Location & dimensions R.R. roads, impounds, fences	●	●	●		
	Well-developed site plot plan & topographical map	●	●			
	Well-developed site facilities	●				
Process flowsheet	Rough sketches				●	
	Preliminary			●		
	Engineered	●	●			
Equipment list	Preliminary sizing & material specifications			●	●	
	Engineered specifications	●	●			
	Vessel sheets	●				
	General arrangement					
	(a) Preliminary			●	●	
	(b) Engineered	●				
Building and structures	Approximate sizes & type of construction			●	●	
	Foundation sketches			●	●	
	Architectural & construction			●	●	
	Preliminary structural design			●		
	General arrangement & elevations	●	●			
	Detailed drawings	●				

Cost-estimating information guide (2/3)

	Required information	Detailed estimate ± 5% range	Definitive estimate ± 10% range	Preliminary estimate ± 20% range	Study estimate ± 30% range	Order-of-magnitude estimate > ± 30% range
Utility requirements	Rough quantities (steam, water, electricity, etc.)				●	
	Preliminary heat balance			●		
	Preliminary flowsheets			●		
	Engineered heat balance	●	●			
	Engineered flowsheet	●				
	Well-developed drawings	●				
Piping	Preliminary flowsheet & specifications			●	●	
	Engineered flowsheet		●			
	Piping layouts & schedules	●	●			
Insulation	Rough specifications			●		
	Preliminary list of equipment & piping to be insulated			●		
	Insulation specifications & schedules	●	●			
	Well-developed drawings or specifications	●	●			
Instrumentation	Preliminary instrument list			●		
	Engineered list & flowsheet	●				
	Well-developed drawings	●				

Cost-estimating information guide (3/3)

	Required information	Detailed estimate ± 5% range	Definitive estimate ± 10% range	Preliminary estimate ± 20% range	Study estimate ± 30% range	Order-of-magnitude estimate > ± 30% range
Electrical	Preliminary motor list—approximate sizes			●	●	
	Engineered list & sizes	●	●			
	Substations, number & sizes, specifications	●	●	●		
	Distribution specifications	●				
	Preliminary lighting specifications			●		
	Preliminary interlock, control, & instrument wiring specs.		●			
	Engineered single-line diagrams (power & light)	●	●			
	Well-developed drawings	●				
Worker-hours	Engineering & drafting	●	●	●		
	Labor by craft	●				
	Supervision	●				
Product scope standard process	Product, capacity, location, & site requirements. Utility & service requirements. Building & auxiliary requirements. Raw materials & finished product handling & storage requirements.					●

Cost estimation – Cost Index

$$\text{Present cost} = \text{original cost}\left(\frac{\text{index value at present}}{\text{index value at time original cost was obtained}}\right)$$

❖ Cost indices as annual averages

Year	Marshall and Swift installed-equipment indexes, 1926 = 100		Eng. News-Record construction index			Nelson-Farrar refinery construction index, 1946 = 100	Chemical Engineering plant cost index, 1957D 1959 = 100
	All Industries	Process industry	1913 = 100	1949 = 100	1967 = 100		
1987	814	830	4406	956	410	1121.5	324
1988	852	859.3	4519	980	421	1164.5	343
1989	895	905.6	4615	1001	430	1195.9	355
1990	915.1	929.3	4732	1026	441	1225.7	357.6
1991	930.6	949.9	4835	1049	450	1252.9	361.3
1992	943.1	957.9	4985	1081	464	1277.3	358.2
1993	964.2	971.4	5210	1130	485	1310.8	359.2
1994	993.4	992.8	5408	1173	504	1349.7	368.1
1995	1027.5	1029.0	5471	1187	509	1392.1	381.1
1996	1039.1	1048.5	5620	1219	523	1418.9	381.7
1997	1056.8	1063.7	5825	1264	542	1449.2	386.5
1998	1061.9	1077.1	5920	1284	551	1477.6	389.5
1999	1068.3	1081.9	6060	1315	564	1497.2	390.6
2000	1089.0	1097.7	6221	1350	579	1542.7	394.1
2001	1093.9	1106.9	6342	1376	591	1579.7	394.3
2002	1102.5‡	1116.9‡	6490‡	1408†	604‡	1599.2†	390.4†§

All costs presented in this text and in the McGraw-Hill website are based on this value for January 2002, obtained from the *Chemical Engineering* index unless otherwise indicated. The website provides the corresponding mathematical cost relationships for all the graphical cost data presented in the text.
†Projected.
§Calculated with revised index; see *Chem. Eng.*, **109:** 62 (2002).

Typical percentages of FCI

Component	Range of FCI, %
Direct costs	
Purchased equipment	15–40
Purchased-equipment installation	6–14
Instrumentation and controls (installed)	2–12
Piping (installed)	4–17
Electrical systems (installed)	2–10
Buildings (including services)	2–18
Yard improvements	2–5
Service facilities (installed)	8–30
Land	1–2
Indirect costs	
Engineering and supervision	4–20
Construction expenses	4–17
Legal expenses	1–3
Contractor's fee	2–6
Contingency	5–15

- The most accurate method for determining process equipment costs is to obtain firm bids from fabricators or suppliers

Estimation for Equipment costs by scaling

$$\text{Cost of equipment } a = (\text{cost of equipment } b) X^{0.6}$$

✓ **The six-tenth factor rule**

(Rule-of-thumb)

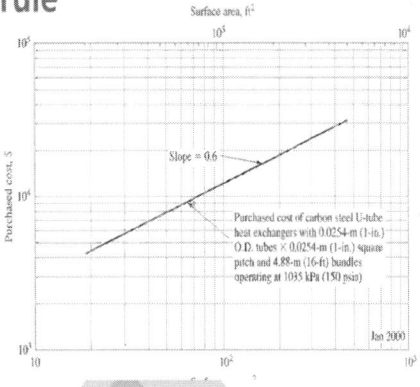

Typical exponents for equipment costs

Equipment	Size range	Exponent
Blender, double cone rotary, carbon steel (c.s.)	1.4–7.1 m³ (50–250 ft³)	0.49
Blower, centrifugal	0.5–4.7 m³/s (10³–10⁴ ft³/min)	0.59
Centrifuge, solid bowl, c.s.	7.5–75 kW (10–10² hp) drive	0.67
Crystallizer, vacuum batch, c.s.	15–200 m³ (500–7000 ft³)	0.37
Compressor, reciprocating, air-cooled, two-stage, 1035-kPa discharge	0.005–0.19 m³ (10–400 ft³/min)	0.69
Compressor, rotary, single-stage, sliding vane, 1035-kPa discharge	0.05–0.5 m³/s (10²–10³ ft³/min)	0.79
Dryer, drum, single vacuum	1–10 m² (10–10² ft²)	0.76
Dryer, drum, single atmospheric	1–10 m² (10–10² ft²)	0.40
Evaporator (installed), horizontal tank	10–1000 m² (10²–10⁴ ft²)	0.54
Fan, centrifugal	0.5–5 m³/s (10³–10⁴ ft³/min)	0.44
Fan, centrifugal	10–35 m³/s (2×10⁴–7×10⁴ ft³/min)	1.17
Heat exchanger, shell-and-tube, floating head, c.s.	10–40 m² (100–400 ft²)	0.60
Heat exchanger, shell-and-tube, fixed sheet, c.s.	10–40 m² (100–400 ft²)	0.44
Kettle, cast-iron, jacketed	1–3 m³ (250–800 gal)	0.27
Kettle, glass-lined, jacketed	0.8–3 m³ (200–800 gal)	0.31
Motor, squirrel cage, induction, 440-V, explosion-proof	4–15 kW (5–20 hp)	0.69
Motor, squirrel cage, induction, 440-V, explosion-proof	15–150 kW (20–200 hp)	0.99
Pump, reciprocating, horizontal cast-iron (includes motor)	1×10⁻⁴–6×10⁻³ m³/s (2–100 gpm)	0.34
Pump, centrifugal, horizontal, cast steel (includes motor)	4–40 m³/s·kPa (10⁴–10⁵ gpm-psi)	0.33
Reactor, glass-lined, jacketed (without drive)	0.2–2.2 m³ (50–600 gal)	0.54
Reactor, stainless steel, 2070-kPa	0.4–4.0 m³ (10²–10³ gal)	0.56
Separator, centrifugal, c.s.	1.5–7 m³ (50–250 ft³)	0.49
Tank, flat head, c.s.	0.4–40 m³ (10²–10⁴ gal)	0.57
Tank, c.s., glass-lined	0.4–4.0 m³ (10²–10³ gal)	0.49
Tower, c.s.	5×10³–10⁶ kg (10⁴–2×10⁶ lb)	0.62
Tray, bubble cap, c.s.	1–3 m (3–10 ft) diameter	1.20
Tray, sieve, c.s.	1–3 m (3–10 ft) diameter	0.86

Cost curve method

> The capital cost of plant can be related to capacity

$$C_2 = C_1 \left(\frac{S_2}{S_1}\right)^n$$

where C_2 = ISBL capital cost of the plant with capacity S_2
C_1 = ISBL capital cost of the plant with capacity S_1

- ✓ n = 0.8 ~ 0.9, for lot of mechanical work or gas compression
   (e.g., methanol, paper pulping, solids-handling plants)
- ✓ n = 0.7, for typical petrochemical processes
- ✓ n = 0.4 ~ 0.5, for small-scale, highly-instrumented processed
   (such as specialty chemical or pharmaceutical manufacture)

❖ For the whole chemical industry n = 0.6

Cost curve method - Economy of scale

$$C_2 = C_1 \left(\frac{S_2}{S_1}\right)^n \implies C_2 = \frac{C_1}{S_1^n} \times S_2^n = a S_2^n \implies \boxed{\frac{C_2}{S_2} = a S_2^{n-1}}$$

- Process Cost Correlation (Examples)

Process	Licensor	Units	Capacity S_{lower}	S_{upper}	a	n
ABS Resin (15% Rubber) by emulsion polymerization	Generic	MMlb/y	50	300	12.146	0.6
Acetic Acid by Cativa process	BP	MMlb/y	500	2,000	3.474	0.6
Acetic Acid by Low Water Methanol Carbonylation	Celanese	MMlb/y	500	2,000	2.772	0.6
Butene-1 by Alphabutol ethylene dimerization	Axens	tpy	5,000	30,000	0.0251	0.6
Butene-1 by BP Process	BP	tpy	20,000	80,000	0.169	0.6
Cyclic Olefin Copolymer by Mitsui Process	Mitsui	MMlb/y	60	120	12.243	0.6
Fischer Tropsch process	ExxonMobil	tpy	200,000	700,000	0.476	0.6
Propylene by Oleflex™ process	UOP	tpy	150,000	350,000	0.0943	0.6
Vinyl acetate by Celanese VAntage Process	Celanese	MMlb/y	300	800	6.647	0.6

Step count method

➢ Plant cost correlates with the number of process step
(for primarily processing liquids and solids)

$$Q \geq 60{,}000: \quad C = 4320\,N \left(\frac{Q}{s}\right)^{0.675}$$

$$Q < 60{,}000: \quad C = 380{,}000\,N \left(\frac{Q}{s}\right)^{0.3}$$

where C = ISBL capital cost in U.S. dollars, U.S. Gulf Coast, Jan. 2010 basis
Q = plant capacity in metric tons per year
s = reactor conversion (= mass of desired product per mass fed to the reactor)
N = number of functional units

A functional unit includes all the equipment and ancillaries needed for a significant process step or function, such as a reaction, separation, or other major unit operation. Pumping and heat exchange are not normally considered as functional units unless they have substantial cost, for example, compressors, refrigeration systems, or process furnaces.

Capital cost data for chemical and petrochemical plants (2000)[+]

Product or process	Process	Typical plant size	Fixed-capital investment, million $	Power factor x for specied process plant
		10^3 kg/yr (10^3 ton/yr)		
Acetic acid	CH_3OH and CO—catalytic	9×10^3 (10)	8	0.68
Acetone	Propylene-copper chloride catalyst	9×10^4 (100)	33	0.45
Ammonia	Steam reforming	9×10^4 (100)	29	0.53
Ammonium nitrate	Ammonia and nitric acid	9×10^4 (100)	6	0.65
Butanol	Propylene, CO, and H_2O—catalytic	4.5×10^4 (50)	48	0.40
Chlorine	Electrolysis of NaCl	4.5×10^4 (50)	33	0.45
Ethylene	Refinery gases	4.5×10^4 (50)	16	0.83
Ethylene oxide	Ethylene—catalytic	4.5×10^4 (50)	59	0.78
Formaldehyde (37%)	Methanol—catalytic	9×10^3 (10)	19	0.55
Glycol	Ethylene and chlorine	4.5×10^3 (5)	18	0.75
Hydrofluoric acid	Hydrogen fluoride and H_2O	9×10^3 (10)	10	0.68
Methanol	CO_2, natural gas, and steam	5.5×10^4 (60)	15	0.60
Nitric acid (high-strength)	Ammonia—catalytic	9×10^4 (100)	8	0.60
Phosphoric acid	Calcium phosphate and H_2SO_4	4.5×10^3 (5)	4	0.60
Polyethylene (high-density)	Ethylene—catalytic	4.5×10^3 (5)	19	0.65
Propylene	Refinery gases	9×10^3 (10)	4	0.70
Sulfuric acid	Sulfur—contact catalytic	9×10^4 (100)	8	0.65
Urea	Ammonia and CO_2	5.5×10^4 (60)	10	0.70
		10^3 m^3/day (10^3 bbl/day)		
Alkylation (H_2SO_4)	Catalytic	1.6 (10)	23	0.60
Coking (delayed)	Thermal	1.6 (10)	31	0.38
Coking (fluid)	Thermal	1.6 (10)	19	0.42
Cracking (fluid)	Catalytic	1.6 (10)	19	0.70
Cracking	Thermal	1.6 (10)	6	0.70
Distillation (atm.)	65% vaporized	16 (100)	38	0.90
Distillation (vac.)	65% vaporized	16 (100)	23	0.70
Hydrotreating	Catalytic desulfurization	1.6 (10)	3.5	0.65
Reforming	Catalytic	1.6 (10)	34	0.60
Polymerization	Catalytic	1.6 (10)	6	0.58

[†] Adapted from K. M. Guthrie, *Chem. Eng.*, 77(13): 140 (1970), and K. M. Guthrie, *Process Plant Estimating, Evaluation, and Control*, Craftsman Book Company of America, Solana Beach, CA, 1974. See also J. E. Haselbarth, *Chem. Eng.*, 74(25): 214 (1967), and D. B. Drayer, *Petro. Chem. Eng.*, 42(5): 10 (1970).
[‡] These power factors apply within roughly a 3-fold ratio extending either way from the plant size as given.

Lang factor method

> Lang(1948) proposed that the ISBL fixed capital cost of plant is given as a function of the total purchased equipment cost

$$C = F\left(\sum C_e\right)$$

where C = total plant ISBL capital cost (including engineering costs)
ΣC_e = total delivered cost of all the major equipment items: reactors, tanks, columns, heat exchangers, furnaces, etc.
F = an installation factor, later widely known as a Lang factor

- Lang factor (based on 1940s economics)

  F = 3.1 for solids processing plant
  F = 4.74 for fluids processing plant
  F = 3.63 for mixed fluids-solids processing plant

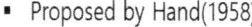

- Proposed by Hand(1958)

Equipment Type	Installation Factor
Compressors	2.5
Distillation columns	4
Fired heaters	2
Heat exchangers	3.5
Instruments	4
Miscellaneous equipment	2.5
Pressure vessels	4
Pumps	4

Location factors

> Local fabrication and construction infrastructure

> Local labor availability and cost

> Costs of shipping or transporting equipment to site

> Import duties or other local tariffs

> Currency exchange rates

Location factors (based on US, 2003)

Country	Region	Location Factor
United States	Gulf Coast	1.00
	East Coast	1.04
	West Coast	1.07
	Midwest	1.02
Canada	Ontario	1.00
	Fort McMurray	1.60
Mexico		1.03
Brazil		1.14
China	imported	1.12
	indigenous	0.61
Japan		1.26
SE Asia		1.12
Australia		1.21
India		1.02
Middle East		1.07
France		1.13
Germany		1.11
Italy		1.14
Netherlands		1.19
Russia		1.53
United Kingdom		1.02

Estimating Offsite Capital Costs

- Guidelines for estimation approximate OSBL costs as a percentage of ISBL cost

	Site Condition		
Process Complexity	Existing: Underused	Existing: Tight Capacity	New Site
Typical large-volume chemical	30%	40%	40%
Low-volume specialty chemical	20%	40%	50%
High solids-handling requirement	40%	50%	100%

생산비용 및 매출 추정

Production cost

➢ Variable production cost

- Raw material
- Utilities
- Consumable cost
- Waste disposal

➢ Fixed production cost

- Labor cost
- Maintenance cost
- Depreciation cost
- Land, Rent & Insurance
- Interest Payments
- License fee and Royalties
- Corporate Overhead Charges
  (R&D, Selling & Marketing,
  General & Administrative cost)

Production cost – Labor cost

> Assessing the minimum number of shift positions

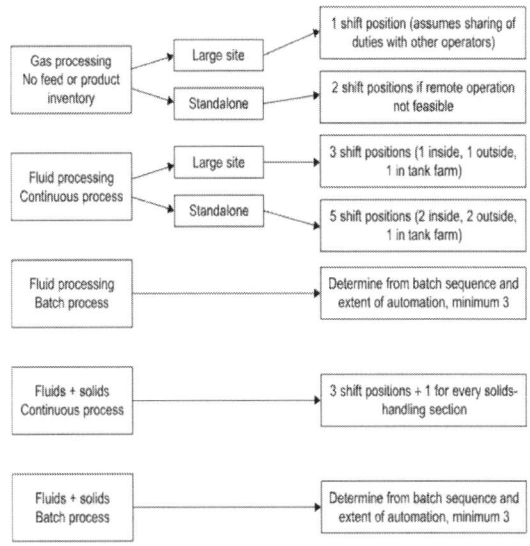

Labor requirements (1/2)

> Typical requirements for process equipment

Type of equipment	Workers/unit/shift
Blowers and compressors	0.1–0.2
Centrifugal separator	0.25–0.50
Crystallizer, mechanical	0.16
Dryer, rotary	0.5
Dryer, spray	1.0
Dryer, tray	0.5
Evaporator	0.25
Filter, vacuum	0.125–0.25
Filter, plate and frame	1.0
Filter, rotary and belt	0.1
Heat exchangers	0.1
Process vessels, towers (including auxiliary pumps and exchangers)	0.2–0.5
Reactor, batch	1.0
Reactor, continuous	0.5

†For expanded process equipment labor requirements see G. D. Ulrich, *A Guide to Chemical Engineering Process Design and Economics*, J. Wiley, New York, 1984.

Labor requirements (2/2)

> Operating labor requirements

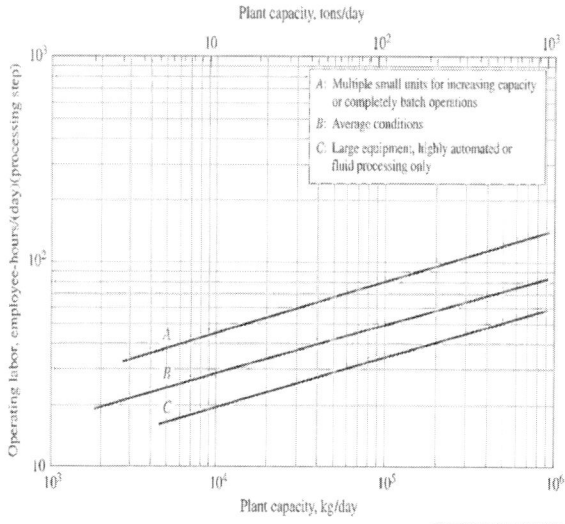

Cost for maintenance and repairs

Type of operation	Maintenance cost as percentage of fixed-capital investment (on annual basis)		
	Wages	Materials	Total
Simple chemical processes	1–3	1–3	2–6
Average processes with normal operating conditions	2–4	3–5	5–9
Complicated processes, severe corrosion operating conditions, or extensive instrumentation	3–5	4–6	7–11

Production Cost Sheet (Example-1/3)

Company Name	Project Name	Adipic acid from phenol					
Address	Project Number					Sheet	1
COST OF PRODUCTION	REV	DATE	BY	APVD	REV	DATE	BY	APVD
Adipic Acid from Phenol	1	1.1.07	GPT				

Form XXXXX-YY-ZZ

Owner's Name		Capital Cost Basis Year	2006	
Plant Location	Northeast Asia	Units	Metric	
Case Description		On Stream	8,000 hr/yr	333.33 day/yr

YIELD ESTIMATE | CAPITAL COSTS

Yield information taken from ChemSystems PERP report 98/99-3, Adipic Acid, p. 89
Yields input for phenol, nitric acid, hydrogen, off-gas, utilities and consumables

Scale of production set to 400 t/y = 880 MMlb/yr

	$MM
ISBL Capital Cost	206.5
OSBL Capital Cost	82.6
Engineering Costs	28.9
Contingency	43.4
Total Fixed Capital Cost	361.3
Working Capital	59.5

REVENUES AND RAW MATERIAL COSTS
MASS BALANCE   MB closure   101%

Key Products	Units	Units/Unit product	Units/yr	Price $/unit	$MM/yr	$/unit main product
Adipic acid	t	1	400,000	1400	560.00	1400.00
Total Key Product Revenues (REV)	t	1	400,000		560.00	1400.00

By-products & Waste Streams						
Nitrous oxide (vented)	t		100,261	0	0.00	0.00
Off-gas	t	0.00417	1,670	700	1.17	2.92
Organic Waste (Fuel value)	t	0.03072	12,286	300	3.69	9.22
Aqueous Waste	t		273,440	-1.5	-0.41	-1.03
Total Byproducts and Wastes (BP)	t	0.0348939	387,659		4.44	11.11

Raw Materials						
Phenol	t	0.71572	286,288	1000	286.29	715.72
Nitric acid 60% (100% basis)	t	0.71778	287,112	380	109.10	272.76
water with nitric acid	t		191,408	0	0.00	0.00
Hydrogen, 99%	t	0.0351	14,040	1100	15.44	38.61
Total Raw Materials (RM)	t	1	778,848		410.83	1027.09

Gross Margin (GM = REV + BP – RM)   153.61   384.03

Production Cost Sheet (Example-2/3)

(Cont'd)

CONSUMABLES

	Units	Units/Unit product	Units/yr	Price $/unit	$MM/yr	$/unit product
Various catalyst and chemicals	kg	32.85	13,138,263	1.00	13.14	32.85
Other	kg	0	0	0.00	0.00	0.00
Total Consumables (CONS)					13.14	32.85

UTILITIES

	Units	Units/Unit product	Units/yr	Price $/unit	$MM/yr	$/unit product
Electric	kWh	206.0	10,300	0.05	4.120	10.30
HP Stream	t	0.4	15	14.30	2.002	5.01
MP Stream	t	7.6	382	12.00	36.624	91.56
LP Stream	t	0.0	0	8.90	0.000	0.00
Boiler Feed	t	0.3	17	1.10	0.145	0.36
Condensate	t	0.0	0	0.80	0.000	0.00
Cooling Water	t	463.0	23,150	0.024	4.445	11.11
Fuel Fired	GJ	0.0	0	5.00	0.000	0.00
Total Utilities (UTS)					47.336	118.340

Variable Cost of Production (VCOP = RM – BP + CONS + UTS)   466.86   1167.16

FIXED OPERATING COSTS

			$MM/yr	$/unit product
Labor	4.8 Operators per Shift Position			
Number of shift positions	9	30,000 $/yr each	1.30	3.24
Supervisions		25% of Operating Labor	0.32	0.81
Direct Ovhd.		45% of Labor & Superv.	0.73	1.82
Maintenance		3% of ISBL Investment	10.84	27.10
Overhead Expense				
Plant Overhead		65% of Labor & Maint.	8.57	21.43
Tax & Insurance		2% of Fixed Investment	5.42	13.55
Interst on Debit Financing		0% of Fixed Capital	0.00	0.00
		6% of Working Capital	3.57	8.93
Fixed Cost of Production (FCOP)			30.75	76.88

Production Cost Sheet (Example-3/3)

(Cont'd)

ANNUALIZED CAPITAL CHARGES	$MM	Interest Rate	Life (yr)	ACCR	$MM/yr	$/unit product
Fixed Capital Investment	361.303	15%	10	0.199	71.99	179.98
Royalty Amortization	15.000	15%	10	0.199	2.99	7.47
Inventory Amortization						
Catalyst 1	0.000	15%	3	0.438	0.00	0.00
Catalyst 2	0.000	15%	3	0.438	0.00	0.00
Adsorbent 1	0.000	15%	3	0.438	0.00	0.00
Equipment 1	0.000	15%	5	0.298	0.00	0.00
Equipment 2	0.000	15%	5	0.298	0.00	0.00
				Total Annual Capital Charge	74.98	187.45

SUMMARY	$MM/yr	$/unit product
Variable Cost of Production	466.86	1167.16
Fixed Cost of Production	30.75	76.88
Cash Cost of Production	497.61	1244.04
Gross Profit	62.39	155.96
Total Cost of Production	572.59	1431.48

Pricing fundamentals

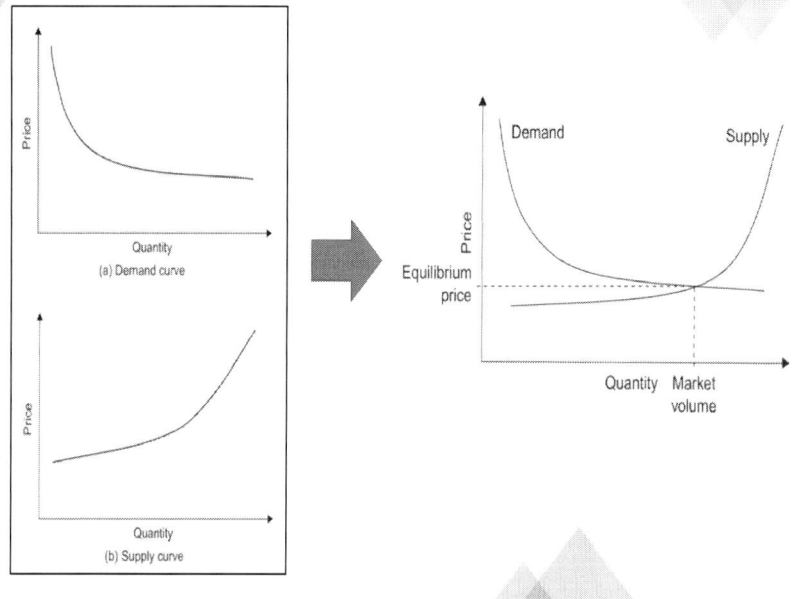

Elasticity of Demand

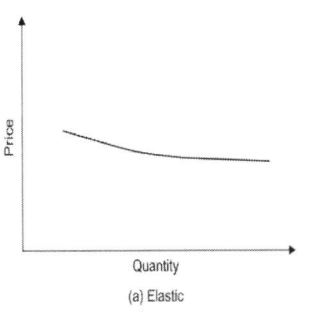
(a) Elastic

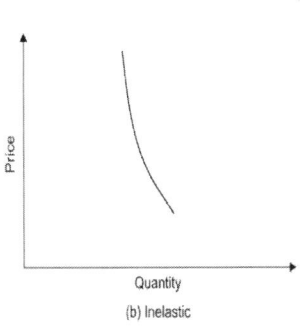
(b) Inelastic

✓ The elasticity of demand depends on
- Availability of substitutes
- Amount of money available to consumers
- Consumer's perception of value or service

✓ The elasticity of supply depends on
- How many producers are able to produce
- How difficult it is to enter the market
- Consumer's perception of value or service

Demand & Supply segmentation

✓ Demand segmentation for air travel

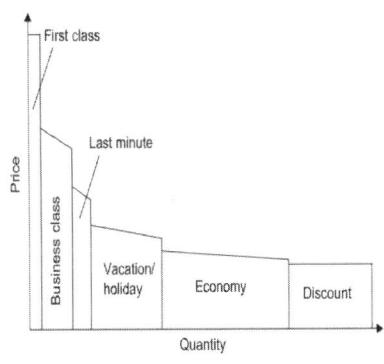

✓ Supply segmentation for commodity chemical

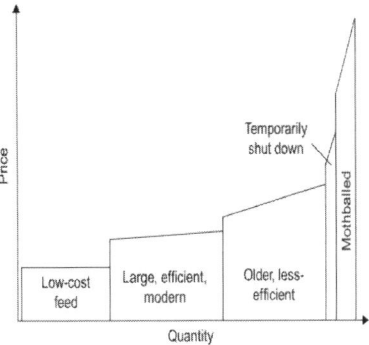

Forecasting prices

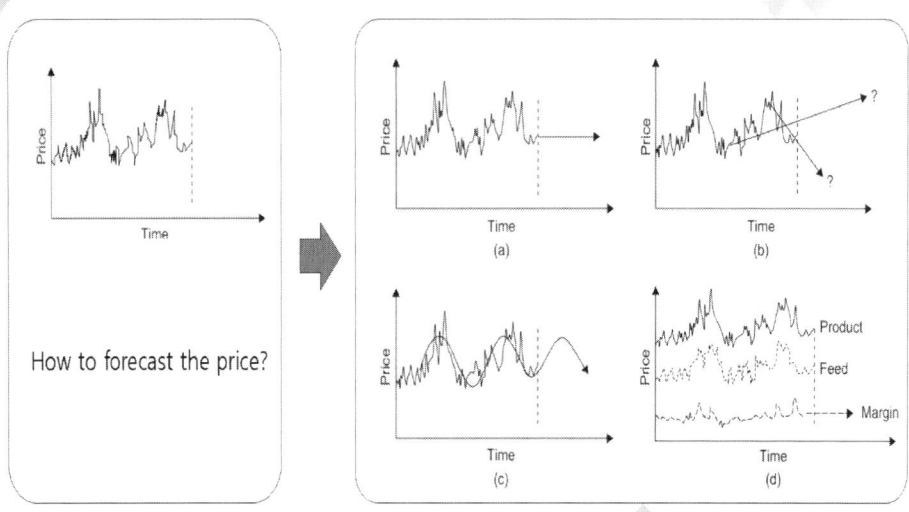

Sources of Price data (for chemicals)

- **Internal Company Forecasts**

- **Trade Journals**
    - ICIS Chemical Business America / Europe / Asia(www.icispricing.com)
    - Oil and Gas Journal

- **Consultants**
    - IHS Markit (former SRI : Chemical Economics Handbook
    - CMAI (Chemical Market Associates Inc.)

- **On-line Supplies and Brokers**
    - www.business.com/directory/chemicals
    - www.molbase.com

- **Reference Books**

Revenues

- The revenues for a project are the income earned from sales of main products and by-products

- The revenues depends on the design basis and is determined by the market size and growth prediction

참고) By-products revenues

- Determining which by-products to recover, purify and sell is usually more difficult than determining the main products

◆ Some Stoichiometric By-products

Feeds	Main Product	By-product
cumene + air	phenol	acetone
propylene + ethylbenzene + air	propylene oxide	styrene
ethylene + chlorine	vinyl chloride monomer	HCl
allyl chloride + HOCl + NaOH	epichlorohydrin	NaCl
methane + steam	hydrogen	carbon dioxide
glucose	ethanol (by fermentation)	carbon dioxide
acetone cyanohydrin + methanol + H_2SO_4	methyl methacrylate	ammonium sulfate
sodium chloride + electricity	chlorine	sodium hydroxide

참고) By-products revenues

> Potentially valuable by-products include as follows

- Materials produced in stoichiometric quantities by the reaction that form from the main product
- Components are produced in high yield by side reaction
  → Ex : Propylene, butadiene in NCC to get main ethylene product
- Components are produced in high yield from feed impurities
  → Ex : most Sulfur is produced as a by-products of fuels manufacture
- Components are produced in low yield that have high value
  → Ex : Dicyclopentadiene can be recovered from the products of NCC
- Degraded consumables such as solvents that have reuse value

Margins

> Gross margin = Revenues − Raw materials costs

- ✓ For petrochemicals and fuels : < 10%
- ✓ For food additives, pharmaceuticals : > 40%

> Variable contribution margin
  = Revenues − Variable costs of production

Profits

❖ The cash cost of production (CCOP)

- CCOP = VCOP + FCOP

  Where VCOP = sum of all the variable cost of production − by-product revenues
  FCOP = sum of the fixed costs of production

➢ Gross profit = Main product revenues − CCOP

➢ Net profit = Gross profit − taxes

Summary : Estimation of capital investment cost

The percentages indicated in the following summary of the various costs constituting the capital investment are approximations applicable to ordinary chemical processing plants. It should be realized that the values given vary depending on many factors, such as plant location, type of process, and complexity of instrumentation.

I. **Direct costs** = material and labor involved in actual installation of complete facility (65–85% of fixed-capital investment)
   A. Equipment + installation + instrumentation + piping + electrical + insulation + painting (50–60% of fixed-capital investment)
      1. Purchased equipment (15–40% of fixed-capital investment)
      2. Installation, including insulation and painting (25–55% of purchased-equipment cost)
      3. Instrumentation and controls, installed (8–50% of purchased-equipment cost)
      4. Piping, installed (10–80% of purchased-equipment cost)
      5. Electrical, installed (10–40% of purchased-equipment cost)
   B. Buildings, process, and auxiliary (10–70% of purchased-equipment cost)
   C. Service facilities and yard improvements (40–100% of purchased-equipment cost)
   D. Land (1–2% of fixed-capital investment or 4–8% of purchased-equipment cost)
II. **Indirect costs** = expenses which are not directly involved with material and labor of actual installation of complete facility (15–35% of fixed-capital investment)
   A. Engineering and supervision (5–30% of direct costs)
   B. Legal expenses (1–3% of fixed-capital investment)
   C. Construction expense and contractor's fee (10–20% of fixed-capital investment)
   D. Contingency (5–15% of fixed-capital investment)
III. **Fixed-capital investment** = direct costs + indirect costs
IV. **Working capital** (10–20% of total capital investment)
V. **Total capital investment** = fixed-capital investment + working capital

Summary : Estimation of total production cost (1/2)

The percentages indicated in the following summary of the various costs involved in the complete operation of manufacturing plants are approximations applicable to ordinary chemical processing plants. It should be realized that the values given vary depending on many factors, such as plant location, type of process, and company policies.

I. **Manufacturing cost** = direct production costs + fixed charges + plant overhead costs
   A. Direct production costs (about 66% of total product cost)
      1. Raw materials (10–80% of total product cost)
      2. Operating labor (10–20% of total product cost)
      3. Direct supervisory and clerical labor (10–20% of operating labor)
      4. Utilities (10–20% of total product cost)
      5. Maintenance and repairs (2–10% of fixed-capital investment)
      6. Operating supplies (10–20% of maintenance and repair costs, or 0.5–1% of fixed-capital investment)
      7. Laboratory charges (10–20% of operating labor)
      8. Patents and royalties (0–6% of total product cost)
   B. Fixed charges (10–20% of total product cost)
      1. Depreciation (depends on method of calculation)
      2. Local taxes (1–4% of fixed-capital investment)
      3. Insurance (0.4–1% of fixed-capital investment)
      4. Rent (8–12% of value of rented land and buildings)
      5. Financing (interest) (0–10% of total capital investment)
   C. Plant overhead costs (50–70% of cost for operating labor, supervision, and maintenance; or 5–15% of total product cost) include costs for the following: general plant upkeep and overhead, payroll overhead, packaging, medical services, safety and protection, restaurants, recreation, salvage, laboratories, and storage facilities

Summary : Estimation of total production cost (2/2)

The percentages indicated in the following summary of the various costs involved in the complete operation of manufacturing plants are approximations applicable to ordinary chemical processing plants. It should be realized that the values given vary depending on many factors, such as plant location, type of process, and company policies.

(Continued)

II. **General expenses** = administrative costs + distribution and selling costs + research and development costs (15–25% of the total product cost)
   A. Administrative costs (about 20% of costs of operating labor, supervision, and maintenance; or 2–5% of total product cost) include costs for executive salaries, clerical wages, computer support, legal fees, office supplies, and communications
   B. Distribution and marketing costs (2–20% of total product cost) include costs for sales offices, salespeople, shipping, and advertising
   C. Research and development costs (2–5% of every sales dollar, or about 5% of total product cost)

III. **Total product cost** = manufacturing cost + general expenses

IV. **Gross earnings cost** (gross earnings = total income − total product cost; amount of gross earnings cost depends on amount of gross earnings for entire company and income tax regulations; a general range for gross earnings cost is 15–40% of gross earnings)

제14장
생산비용 및 매출 추정

1) 생산비용 및 매출 추정

2) 투자 자금, 이자, 돈의 시간적 가치 및 수익성

Capital Cost Investment

✓ Economic definition of **_capital_** is "stock of accumulated wealth"

✓ **_Investment_** is to promote the production of other goods
  ... based on the view of obtaining _income_ or _profit_

❖ Fixed Capital Investment (FCI, 고정자본 투자비)
  - to supply the required manufacturing and plant facilities

❖ Working Capital (WC, 운전자본)
  - necessary cost for the operation of the plant

◆ Total Investment Cost (TCI, 총 자본 투자비) = FCI + WC

Components of Capital cost

❖ Fixed Capital

The total cost of designing, constructing, and installing a plant and the associated modification needed to prepare the plant site

- ISBL
- OSBL
- Engineering and construction cost
- Contingency charges

* ISBL : Inside battery limit
* OSBL : Offsite battery limit

❖ Working Capital

The additional money to start the plant up and keep it running. It is the money that is tied up in sustaining plant operation

- Value of raw material inventory
- Value of product inventory
- Cash for cost of production (1wk)
- Accounts receivable – products shipped but not paid (1month)
- Credit for accounts payable – feedstocks, solvents etc.
- Spare parts inventory
  -1~2% ISBL+ OSBL

생산비용 및 매출 추정

Production cost

➢ Variable production cost

- Raw material
- Utilities
- Consumable cost
- Waste disposal

➢ Fixed production cost

- Labor cost
- Maintenance cost
- Depreciation cost
- Land, Rent & Insurance
- Interest Payments
- License fee and Royalties
- Corporate Overhead Charges

  (R&D, Selling & Marketing, General & Administrative cost)

Production cost – Labor cost

➢ Assessing the minimum number of shift positions

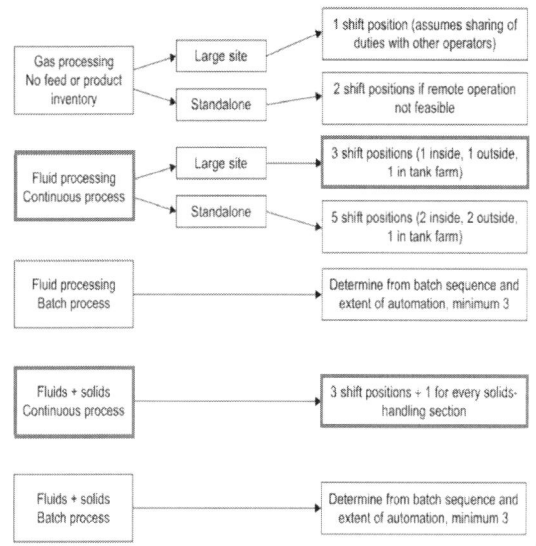

Labor requirements (1/2)

> Typical requirements for process equipment

Type of equipment	Workers/unit/shift
Blowers and compressors	0.1–0.2
Centrifugal separator	0.25–0.50
Crystallizer, mechanical	0.16
Dryer, rotary	0.5
Dryer, spray	1.0
Dryer, tray	0.5
Evaporator	0.25
Filter, vacuum	0.125–0.25
Filter, plate and frame	1.0
Filter, rotary and belt	0.1
Heat exchangers	0.1
Process vessels, towers (including auxiliary pumps and exchangers)	0.2–0.5
Reactor, batch	1.0
Reactor, continuous	0.5

†For expanded process equipment labor requirements see G. D. Ulrich, *A Guide to Chemical Engineering Process Design and Economics*, J. Wiley, New York, 1984.

Labor requirements (2/2)

> Operating labor requirements

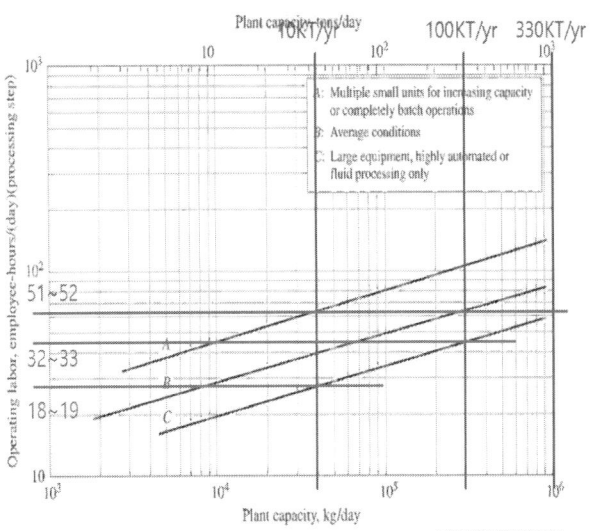

Cost for maintenance and repairs

Type of operation	Maintenance cost as percentage of fixed-capital investment (on annual basis)		
	Wages	Materials	Total
Simple chemical processes	1–3	1–3	2–6
Average processes with normal operating conditions	2–4	3–5	5–9
Complicated processes, severe corrosion operating conditions, or extensive instrumentation	3–5	4–6	7–11

Production Cost Sheet (Example-1/3)

Company Name				Project Name	Adipic acid from phenol				
Address				Project Number				Sheet	1
COST OF PRODUCTION				REV	DATE	BY	APVD	REV	DATE	BY	APVD
Adipic Acid from Phenol				1	1.1.07	GPT			

Form XXXXX-YY-ZZ

Owner's Name		Capital Cost Basis Year 2006	
Plant Location	Northeast Asia	Units	Metric
Case Description		On Stream	8,000 hr/yr 333.33 day/yr

YIELD ESTIMATE

Yield information taken from ChemSystems PERP report 98/99-3 Adipic Acid, p. 89
Yields input for phenol, nitric acid, hydrogen, off-gas, utilities and consumables

Scale of production set to 400 t/y = 880 MMlb/yr

CAPITAL COSTS

	$MM
ISBL Capital Cost	206.5
OSBL Capital Cost	82.6
Engineering Costs	28.9
Contingency	43.4
Total Fixed Capital Cost	361.3
Working Capital	59.5

REVENUES AND RAW MATERIAL COSTS

MASS BALANCE   MB closure   101%

Key Products	Units	Units/Unit product	Units/yr	Price $/unit	$MM/yr	$/unit main product
Adipic acid	t	1	400,000	1400	560.00	1400.00
Total Key Product Revenues (REV)	t	1	400,000		560.00	1400.00
By-products & Waste Streams						
Nitrous oxide (vented)	t		100,261	0	0.00	0.00
Off-gas	t	0.00417	1,670	700	1.17	2.92
Organic Waste (Fuel value)	t	0.03072	12,288	300	3.69	9.22
Aqueous Waste	t		273,440	-1.5	-0.41	-1.03
Total Byproducts and Wastes (BP)	t	0.0348939	387,659		4.44	11.11
Raw Materials						
Phenol	t	0.71572	286,268	1000	286.29	715.72
Nitric acid 60% (100% basis)	t	0.71778	287,112	380	109.10	272.76
water with nitric acid	t		191,408	0	0.00	0.00
Hydrogen 99%	t	0.0351	14,040	1100	15.44	38.61
Total Raw Materials (RM)		1	778,848		410.83	1027.09
			Gross Margin (GM = REV + BP – RM)		153.61	384.03

Production Cost Sheet (Example-2/3)

(Cont'd)

CONSUMABLES

	Units	Units/Unit product	Units/yr	Price $/unit	$MM/yr	$/unit product
Various catalyst and chemicals	kg	32.85	13,138,263	1.00	13.14	32.85
Other	kg	0	0	0.00	0.00	0.00
Total Consumables (CONS)					13.14	32.85

UTILITIES

	Units	Units/Unit product	Units/yr	Price $/unit	$MM/yr	$/unit product
Electric	kWh	206.0	10,300	0.05	4.120	10.30
HP Stream	t	0.4	18	14.30	2.002	5.01
MP Stream	t	7.6	382	12.00	36.624	91.56
LP Stream	t	0.0	0	8.90	0.000	0.00
Boiler Feed	t	0.3	17	1.10	0.145	0.36
Condensate	t	0.0	0	0.80	0.000	0.00
Cooling Water	t	463.0	23,150	0.024	4.445	11.11
Fuel Fired	GJ	0.0	0	6.00	0.000	0.00
Total Utilities (UTS)					47.336	118.340
Variable Cost of Production (VCOP = RM − BP + CONS + UTS)					466.86	1167.16

FIXED OPERATING COSTS

			$MM/yr	$/unit product
Labor	4.8 Operators per Shift Position			
Number of shift positions	9	30,000 $/yr each	1.30	3.24
Supervisions		25% of Operating Labor	0.32	0.81
Direct Ovhd.		45% of Labor & Superv.	0.73	1.82
Maintenance		3% of ISBL Investment	10.84	27.10
Overhead Expense				
Plant Overhead		65% of Labor & Maint.	8.57	21.43
Tax & Insurance		2% of Fixed Investment	5.42	13.55
Interst on Debit Financing		0% of Fixed Capital	0.00	0.00
		6% of Working Capital	3.57	8.93
Fixed Cost of Production (FCOP)			30.75	76.88

Production Cost Sheet (Example-3/3)

(Cont'd)

ANNUAIZED CAPITAL CHARGES

	$MM	Interest Rate	Life (yr)	ACCR	$MM/yr	$/unit product
Fixed Capital Investment	361.303	15%	10	0.199	71.99	179.98
Royalty Amortization	15.000	15%	10	0.199	2.99	7.47
Inventory Amortization						
Catalyst 1	0.000	15%	3	0.438	0.00	0.00
Catalyst 2	0.000	15%	3	0.438	0.00	0.00
Adsorbent 1	0.000	15%	3	0.438	0.00	0.00
Equipment 1	0.000	15%	5	0.298	0.00	0.00
Equipment 2	0.000	15%	5	0.298	0.00	0.00
Total Annual Capital Charge					74.98	187.45

SUMMARY

	$MM/yr	$/unit product
Variable Cost of Production	466.86	1167.16
Fixed Cost of Production	30.75	76.88
Cash Cost of Production	497.61	1244.04
Gross Profit	62.39	155.96
Total Cost of Production	572.59	1431.48

Pricing fundamentals

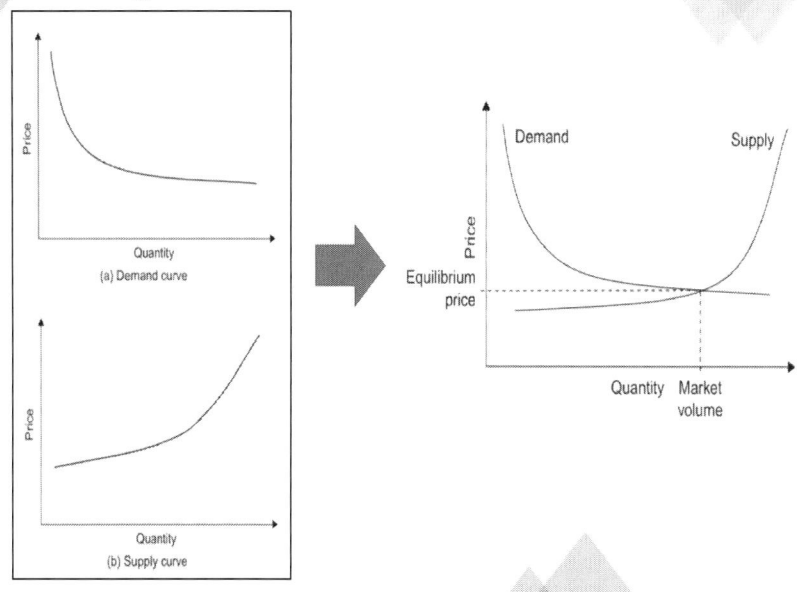

Elasticity of Demand

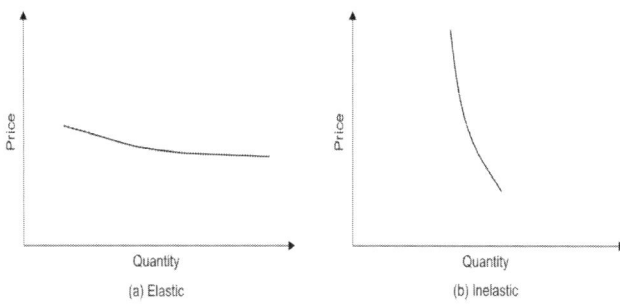

- ✓ The elasticity of demand depends on
  - Availability of substitutes
  - Amount of money available to consumers
  - Consumer's perception of value or service

- ✓ The elasticity of supply depends on
  - How many producers are able to produce
  - How difficult it is to enter the market
  - Consumer's perception of value or service

Demand & Supply segmentation

✓ Demand segmentation for air travel

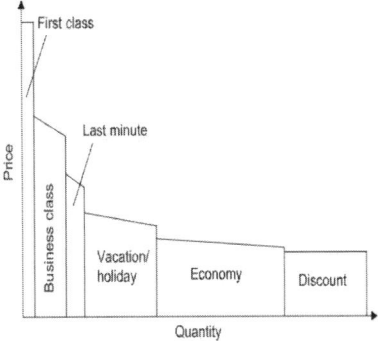

✓ Supply segmentation for commodity chemical

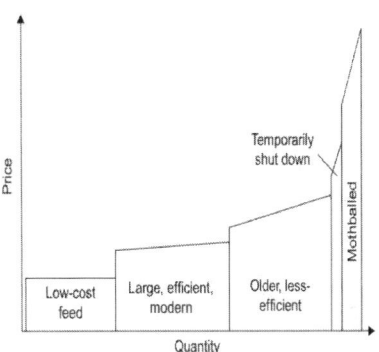

Forecasting prices

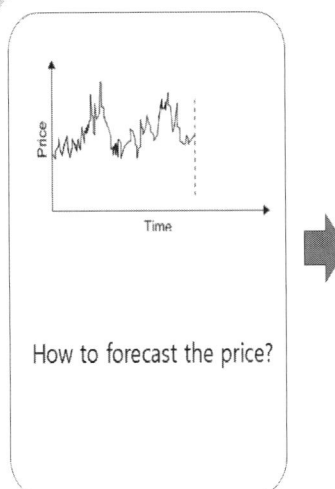

How to forecast the price?

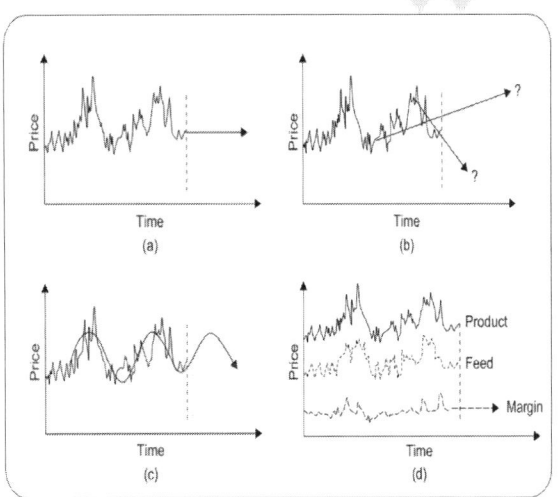

Sources of Price data (for chemicals)

- Internal Company Forecasts

- Trade Journals
    - ICIS Chemical Business America / Europe / Asia(www.icispricing.com)
    - Oil and Gas Journal

- Consultants
    - IHS Markit (former SRI : Chemical Economics Handbook
    - CMAI (Chemical Market Associates Inc.)

- On-line Supplies and Brokers
    - www.business.com/directory/chemicals
    - www.molbase.com

- Reference Books

Revenues

- The revenues for a project are the income earned from sales of main products and by-products

- The revenues depends on the design basis and is determined by the market size and growth prediction

참고) By-products revenues

> Determining which by-products to recover, purify and sell is usually more difficult than determining the main products

♦ Some Stoichiometric By-products

Feeds	Main Product	By-product
cumene + air	phenol	acetone
propylene + ethylbenzene + air	propylene oxide	styrene
ethylene + chlorine	vinyl chloride monomer	HCl
allyl chloride + HOCl + NaOH	epichlorohydrin	NaCl
methane + steam	hydrogen	carbon dioxide
glucose	ethanol (by fermentation)	carbon dioxide
acetone cyanohydrin + methanol + H_2SO_4	methyl methacrylate	ammonium sulfate
sodium chloride + electricity	chlorine	sodium hydroxide

참고) By-products revenues

> Potentially valuable by-products include as follows

- Materials produced in stoichiometric quantities by the reaction that form from the main product

- Components are produced in high yield by side reaction
  → Ex : Propylene, butadiene in NCC to get main ethylene product

- Components are produced in high yield from feed impurities
  → Ex : most Sulfur is produced as a by-products of fuels manufacture

- Components are produced in low yield that have high value
  → Ex : Dicyclopentadiene can be recovered from the products of NCC

- Degraded consumables such as solvents that have reuse value

Margins

➢ Gross margin = Revenues − Raw materials costs

  ✓ For petrochemicals and fuels : < 10%
  ✓ For food additives, pharmaceuticals : > 40%

➢ Variable contribution margin
      = Revenues − Variable costs of production

Profits

❖ The cash cost of production (CCOP)

  ▪ CCOP = VCOP + FCOP

  Where VCOP = sum of all the variable cost of production − by-product revenues
  FCOP = sum of the fixed costs of production

➢ Gross profit = Main product revenues − CCOP

➢ Net profit = Gross profit − taxes

Summary : Estimation of capital investment cost

The percentages indicated in the following summary of the various costs constituting the capital investment are approximations applicable to ordinary chemical processing plants. It should be realized that the values given vary depending on many factors, such as plant location, type of process, and complexity of instrumentation.

I. **Direct costs** = material and labor involved in actual installation of complete facility (65–85% of fixed-capital investment)
  A. Equipment + installation + instrumentation + piping + electrical + insulation + painting (50–60% of fixed-capital investment)
    1. Purchased equipment (15–40% of fixed-capital investment)
    2. Installation, including insulation and painting (25–55% of purchased-equipment cost)
    3. Instrumentation and controls, installed (8–50% of purchased-equipment cost)
    4. Piping, installed (10–80% of purchased-equipment cost)
    5. Electrical, installed (10–40% of purchased-equipment cost)
  B. Buildings, process, and auxiliary (10–70% of purchased-equipment cost)
  C. Service facilities and yard improvements (40–100% of purchased-equipment cost)
  D. Land (1–2% of fixed-capital investment or 4–8% of purchased-equipment cost)
II. **Indirect costs** = expenses which are not directly involved with material and labor of actual installation of complete facility (15–35% of fixed-capital investment)
  A. Engineering and supervision (5–30% of direct costs)
  B. Legal expenses (1–3% of fixed-capital investment)
  C. Construction expense and contractor's fee (10–20% of fixed-capital investment)
  D. Contingency (5–15% of fixed-capital investment)
III. **Fixed-capital investment** = direct costs + indirect costs
IV. **Working capital** (10–20% of total capital investment)
V. **Total capital investment** = fixed-capital investment + working capital

Summary : Estimation of total production cost (1/2)

The percentages indicated in the following summary of the various costs involved in the complete operation of manufacturing plants are approximations applicable to ordinary chemical processing plants. It should be realized that the values given vary depending on many factors, such as plant location, type of process, and company policies

I. **Manufacturing cost** = direct production costs + fixed charges + plant overhead costs
  A. Direct production costs (about 66% of total product cost)
    1. Raw materials (10–80% of total product cost)
    2. Operating labor (10–20% of total product cost)
    3. Direct supervisory and clerical labor (10–20% of operating labor)
    4. Utilities (10–20% of total product cost)
    5. Maintenance and repairs (2–10% of fixed-capital investment)
    6. Operating supplies (10–20% of maintenance and repair costs, or 0.5–1% of fixed-capital investment)
    7. Laboratory charges (10–20% of operating labor)
    8. Patents and royalties (0–6% of total product cost)
  B. Fixed charges (10–20% of total product cost)
    1. Depreciation (depends on method of calculation
    2. Local taxes (1–4% of fixed-capital investment)
    3. Insurance (0.4–1% of fixed-capital investment)
    4. Rent (8–12% of value of rented land and buildings)
    5. Financing (interest) (0–10% of total capital investment)
  C. Plant overhead costs (50–70% of cost for operating labor, supervision, and maintenance; or 5–15% of total product cost) include costs for the following: general plant upkeep and overhead, payroll overhead, packaging, medical services, safety and protection, restaurants, recreation, salvage, laboratories, and storage facilities

Summary : Estimation of total production cost (2/2)

The percentages indicated in the following summary of the various costs involved in the complete operation of manufacturing plants are approximations applicable to ordinary chemical processing plants. It should be realized that the values given vary depending on many factors, such as plant location, type of process, and company policies.

(Continued)

II. **General expenses** = administrative costs + distribution and selling costs + research and development costs (15–25% of the total product cost)
   A. Administrative costs (about 20% of costs of operating labor, supervision, and maintenance; or 2–5% of total product cost) include costs for executive salaries, clerical wages, computer support, legal fees, office supplies, and communications
   B. Distribution and marketing costs (2–20% of total product cost) include costs for sales offices, salespeople, shipping, and advertising
   C. Research and development costs (2–5% of every sales dollar, or about 5% of total product cost)

III. **Total product cost** = manufacturing cost + general expenses

IV. **Gross earnings cost** (gross earnings = total income − total product cost; amount of gross earnings cost depends on amount of gross earnings for entire company and income tax regulations; a general range for gross earnings cost is 15–40% of gross earnings)

  투자 자금

Capital Cost Investment

✓ Economic definition of **_capital_** is "stock of accumulated wealth"

✓ **_Investment_** is to promote the production of other goods
... based on the view of obtaining _income_ or _profit_

❖ Fixed Capital Investment (FCI, 고정자본 투자비)
   - to supply the required manufacturing and plant facilities

❖ Working Capital (WC, 운전자본)
   - necessary cost for the operation of the plant

◆ Total Investment Cost (TCI, 총 자본 투자비) = FCI + WC

Components of Capital cost

❖ Fixed Capital

The total cost of designing, constructing, and installing a plant and the associated modification needed to prepare the plant site

- ISBL
- OSBL
- Engineering and construction cost
- Contingency charges

* ISBL : Inside battery limit
* OSBL : Offsite battery limit

❖ Working Capital

The additional money to start the plant up and keep it running. It is the money that is tied up in sustaining plant operation

- Value of raw material inventory
- Value of product inventory
- Cash for cost of production (1wk)
- Accounts receivable – products shipped but not paid (1month)
- Credit for accounts payable – feedstocks, solvents etc.
- Spare parts inventory
  -1~2% ISBL+ OSBL

Funds for a proposed project

❖ **The required fund, from where?**

- The Borrowed Fund → What is the interest rate
- The Internal Fund → How to decide,
  what is the best project
  between project A and B

❖ **Time value of money for earnings and investments**

❖ **And then, Taxes for net profit**

참고) 식당 창업 및 운영자금

❖ **창업 시, 기본 고려사항**
- 자가 or 임대
- 식기류, 식재료, 음료 및 기타 설비비
- 직원 채용, 교육 및 인건비
- 개업 전 광고

❖ **현금 유동성**
- 일일 거래 현금 수입 및 지출
- 선수금, 외상 매출, 미지급금 및 외상 매입금 등개업 전 광고
- 재고회전율
- 지불준비금

Source : m.blog.naver.com/PostView.naver?isHttpsRedirect=true&blogId=restartfactory&logNo=221123882268

이자

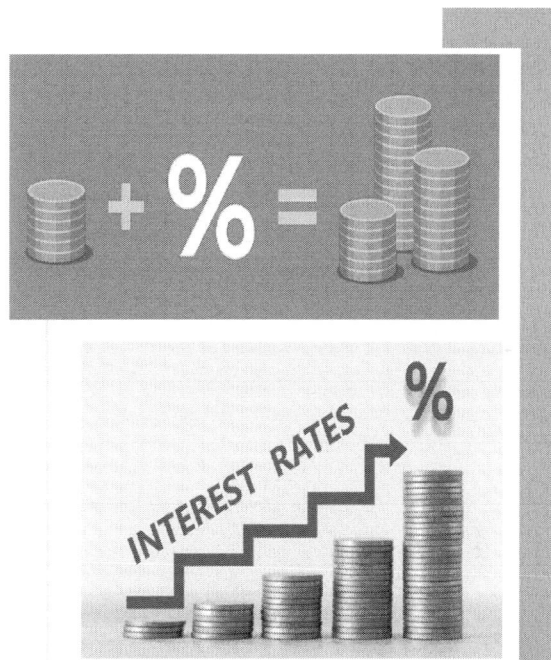

Types of Interest

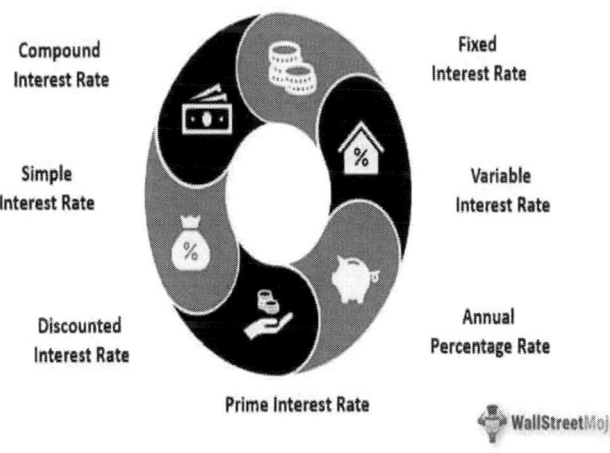

Source : www.wallstreetmojo.com/types-of-interest/

Fixed Interest Rate (고정금리, 1/2)

A fixed interest rate is the most common type of interest rate, which is generally charged to the borrower of the loan by lenders. As the name suggests, the interest rate is fixed throughout the loan's repayment period. It is usually decided on an agreement basis between the lender and the borrower when the loan is granted. This is much easier, and calculations are not at all complex.

- It gives a clear understanding to the lender and the borrower what the exact amount of interest rate obligation is associated with the loan.
- Fixed interest is a rate that does not fluctuate with time or during the loan period. This helps in the accurate estimation of future payments to the borrower.
- Though one drawback of a fixed interest rate is that it can be higher than variable interest rates, it eventually avoids the risk that a loan or mortgage can get costly.

Source : www.wallstreetmojo.com/types-of-interest/

Fixed Interest Rate (고정금리, 2/2)

Example

A fixed interest rate can be a borrower who has taken a home loan from a bank/lender for a sum of $100000 at a 10% interest rate for a period of 15 years. This means the borrower for 15 years must bear 10% of $100000 = $10000 every year as the interest payment. Thus, with the principal amount constantly every year, he has to pay $10000 for 15 years. Thus, we see no change in the rate of interest and the interest amount which the borrower has to repay the bank. Thus, it makes it easy for the borrower to plan his budget accordingly and make the payment.

Source : www.wallstreetmojo.com/types-of-interest/

Variable Interest Rate (변동금리, 1/2)

A variable interest rate is just the opposite of a fixed interest rate. Here the interest rate fluctuates with time. Variable-rate interest is generally linked to the movement of the base level of interest rate, which is also called the prime rate of interest. Borrowers end up on the winning side if the loan has opted on a variable rate of interest basis, and the prime lending rate decreases.

- In this case, the borrowing rate also goes down. This generally happens when the economy is passing through a crisis. On the other hand, if the base interest rate or the prime interest rate rises, the borrower is forced to pay a higher interest rate in such scenarios. Banks will purposely do such to safeguard themselves from interest rates as low as that the borrower ends up giving payments, which are comparatively lesser than the market value of the interest for the loan or debt.

- Similarly, the borrower has an added advantage when the prime rate of interest falls after a loan is approved. The borrower does not have to overpay for the loan with the variable rate assigned to the prime interest rate.

Source : www.wallstreetmojo.com/types-of-interest/

Variable Interest Rate (변동금리, 2/2)

Example

Suppose the borrower is given a home loan for 15 years, and the loan amount sanctioned is $100000 at a 10% interest rate. The contract is set as for the first five years, the borrower will pay a fixed rate of 10 %, i.e., $10000 years, whereas, after five years, the interest rate will be on a variable basis assigned to the prime interest rate or base rate. Now suppose after five years, the prime rate increases, which eventually increases the borrowing rate to 11 %. Thus now the borrower pays $11,000 yearly, whereas if the prime rate falls and the borrowing rate becomes 9%, the borrower in such a scenario saves money and only ends up paying $9,000 yearly.

Source : www.wallstreetmojo.com/types-of-interest/

Annual Percentage Rate

Annual Percentage Rate is very common in credit card companies and credit card mode of payment methodology. Here the annual rate of interest is calculated as the amount of the total sum of interest pending, which is expressed on the total cost of the loan.

- Credit card companies will apply this method when customers carry their balance forward instead of repaying it fully. The calculation of the annual percentage rate is expressed as the prime interest rate; along with this, the margin that the bank or lender charges are added upon.

Example

Suppose we have a credit card with a 24% APR. It means for 12 months, we are charged at a rate of 2% per month. Now all months won't have equal days; thus, APR is further divided by 365 days or 0.065%, which is called the DPR. Thus interest rate finally stands for DPR or the daily rate multiplied by the daily card balance, and then further, this result is multiplied by the number of days in the billing cycle.

Source : www.wallstreetmojo.com/types-of-interest/

참고) Billing Cycle (대금 청구 주기)

The billing cycle is the period between one billing statement and the next billing date that companies generate for their services and products sold to the customers. The customers make payments based on the bill invoices received from suppliers, and this cycle doesn't need to be monthly. It mostly depends on the type of service or goods sold.

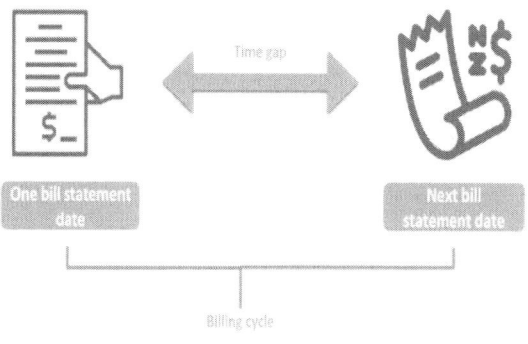

Source : www.wallstreetmojo.com/types-of-interest/

Prime Interest Rate (우대금리, 1/2)

The prime rate is the rate the banks generally give to their favored customers or customers with a very good credit history. This rate is generally lower than the usual lending/borrowing rate. It is generally linked to the Federal Reserve lending rate, the rate at which different banks borrow and lend. But again, not all customers will be able to opt for this loan.

> Prime Rate (also known as Prime Lending Rate, i.e., PLR) is an interest rate offered by banks for their most creditworthy customers. These customers are generally huge corporates with a spotless credit history. These rates are not centralized and can vary from bank to bank.It forms the basis for interest rates on business loans, personal loans, vehicle loans, home loans, mortgages, etc.

Source : www.wallstreetmojo.com/types-of-interest/

Prime Interest Rate (우대금리, 2/2)

Example

Suppose when a big corporation has a regular loan history and very good repayment history, too, with the bank approaching the lender for a short term loan, the bank can arrange for the same at a prime rate and offer it to its customer as a good gesture of relationship.

> Short-term loans are defined as borrowings undertaken for a short period to meet immediate monetary requirements. For example, companies often borrow short-term loans using bank overdrafts to arrange money for working capital requirements.

Source : www.wallstreetmojo.com/types-of-interest/

Discounted Interest Rate

This interest rate does not apply to the common public. This rate is generally applicable for Federal Banks to lend money to other financial institutions on a short-term basis, which can be as short as a single day. Banks may opt for such loans at a discounted rate to cover up their lending capacity, rectify liquidity problems, or prevent a bank from failing in a crisis.

Example

Suppose, at times when the loans/lending becomes more than deposits in a single day, a particular bank may approach the Federal Bank to grant loans at a discounted rate to cover up their liquidity or lending position for the day.

Source : www.wallstreetmojo.com/types-of-interest/

Simple Interest Rate (단리)

Simple Interest is a bank's rate of interest for charging its customers. The calculation is basic and generally expressed as the multiplication of principal, interest rate, and the number of periods.

Example

Suppose a bank is charging a 10% rate of interest on a loan for $1000 for three years; the simple interest calculation stands to be $1000 * 10% *3 = $300.

Source : www.wallstreetmojo.com/types-of-interest/

Compound Interest Rate (복리)

Compound Interest methodology is called interest on interest. Banks generally use the calculation to calculate the bank rates. It is based on two key elements: the interest of the loan and the principal amount. Here banks will first apply the interest amount on the loan balance, and whatever balance is pending will use the same amount to calculate the subsequent year's interest payment.

Example

For example, we have invested in the bank for $1000 at 10% interest. First-year we will earn $100 and second-year the interest rate will be calculated not on $1,000 but on $1,000 + $100 = $1,100. Thus we will earn slightly more than what we would have earned under a simple interest format.

Source : www.wallstreetmojo.com/types-of-interest/

Equations of Interest

If P represents the principal, N the number of time units or interest periods, and i the interest rate based on the length of one interest period, the amount of simple interest I accumulated during N interest periods is

$$I = PiN$$ 　　　(P : Principal, 원금)

$$F = P + I = P(1 + iN)$$

where F is the total amount of principal and accumulated interest at time N.

- Equations for Simple and Compound interest

Period	Principal at start of period	Interest earned during period (i = interest rate based on length of one period)	Compound amount F at end of period
1	P	Pi	$P + Pi = P(1+i)$
2	$P(1+i)$	$P(1+i)(i)$	$P(1+i) + P(1+i)(i) = P(1+i)^2$
3	$P(1+i)^2$	$P(1+i)^2(i)$	$P(1+i)^2 + P(1+i)^2(i) = P(1+i)^3$
N	$P(1+i)^{N-1}$	$P(1+i)^{N-1}(i)$	$P(1+i)^N$

Time basis for Compound interest

❖ Effective annual interest rate
(실질이율)

✓ Or i_{eff}
✓ Allows interest calculations to be made on an annual basis and gives the same result as using the actual compounding periods
✓ Greater than the nominal annual interest rate.
✓ Indicates that the effective interest rate will continue to increase as the number of compounding periods per year increases.
✓ Formulae

$$i_{eff} = \left(1 + \frac{i_{nom}}{m}\right)^m - 1$$

For calculations of year by year

❖ Nominal annual interest rate
(명목이율)

✓ Terms
  ❖ I_{nom} = Nominal annual interest rate
  ❖ m = Compounding periods per year, 12
  ❖ r = Rate per compounding period
✓ Rate per compounding period

$$r = \frac{i_{nom}}{m} \qquad F_1 = P\left(1 + \frac{i_{nom}}{m}\right)^m$$

For calculations of months in a year

Comparison with simple and compound interest

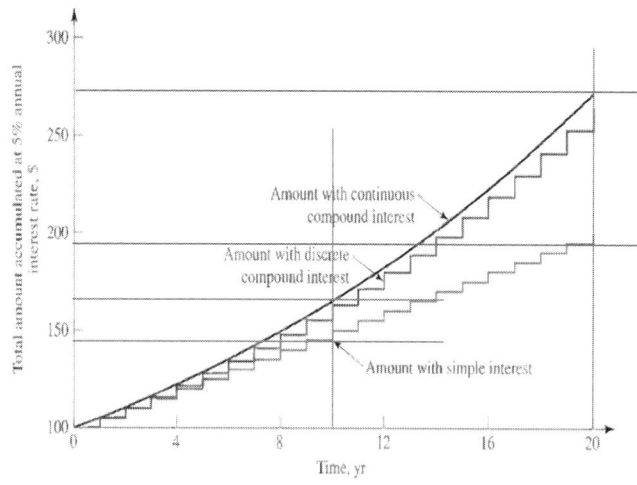

Loan Payments (1/2)

❖ Assume the constant amount and the required interest must be paid at given period

$$L = I_j + p_j \quad \text{Eq-A}$$

where L is the constant payment each period, I_j the jth period interest payment, and p_j the jth period principal payment. The index j begins at 1, because payment is made at the end of the period.

✓ The interest payment (@ effective annual interest rate)

$$\boxed{I_j = iP_{j-1}} \quad \text{Eq-B}$$

where i is the interest rate and P_{j-1} the principal balance after payment $j-1$.

Loan Payments (2/2)

✓ The remaining principal balance after j-1 period

$$\boxed{P_{j-1} = P_0 - \sum_{m=1}^{j-1} p_m} \quad \text{Eq-C}$$

where p_m is the mth principal payment and P_0 the initial amount of the loan.

✓ From eqs.- A, B, and C

$$\boxed{p_j = L - I_j = L - i\left(P_0 - \sum_{m=1}^{j-1} p_m\right)} \quad \text{Eq-D}$$

✓ Finally, we get

$$\boxed{L = \frac{P_0\left[1 + i\sum_{1}^{N}(1+i)^{j-1}\right]}{\sum_{1}^{N}(1+i)^{j-1}}} \quad \text{Eq-E}$$

Ex.) Calculation of Loan Payments

- Loan : USD 100,000
- Interest : 10%/year (Nominal compound interest rate)
- Payment period : 10 years

❖ Determine as follows :

- Constant payment per year,
- Interest and principal paid each period
- Remaining unpaid principal at the end of each period
    - Monthly Loan Payments
    - Yearly Loan Payments

Ex.) Monthly Loan Payments

Month j	$(1+0.1/12)^{j-1}$ Column sum = 204.84498	Constant payment L, $/month, Eq. E	Interest payment, $/month, Eq. B	Principal payment, $/month, Eq. D	Remaining principal, $, Eq. C
0					100,000.00
1	1.0000000	1,321.51	833.33	488.17	99,511.83
2	1.0083333	1,321.51	829.27	492.24	99,019.58
3	1.0167361	1,321.51	825.16	496.34	98,523.24
4	1.0252089	1,321.51	821.03	500.48	98,022.76
5	1.0337523	1,321.51	816.86	504.65	97,518.11
6	1.0423669	1,321.51	812.65	508.86	97,009.25
7	1.0510533	1,321.51	808.41	513.10	96,496.15
8	1.0598121	1,321.51	804.13	517.37	95,978.78
9	1.0686439	1,321.51	799.82	521.68	95,457.10
10	1.0775492	1,321.51	795.48	526.03	94,931.07
11	1.0865288	1,321.51	791.09	530.42	94,400.65
12	1.0955832	1,321.51	786.67	534.84	93,865.82
From here, most results are deleted; only those for every 12th month are shown.					
24	1.2103051	1,321.51	730.67	590.84	87,089.30
36	1.3370398	1,321.51	668.80	652.71	79,603.20
48	1.4770454	1,321.51	600.45	721.06	71,333.20
60	1.6317113	1,321.51	524.95	796.56	62,197.23
72	1.8025728	1,321.51	441.54	879.97	52,104.60
84	1.9913258	1,321.51	349.39	972.11	40,955.15
96	2.1998436	1,321.51	247.60	1,073.91	28,638.19
108	2.4301960	1,321.51	135.15	1,186.36	15,031.50
120	2.6846692	1,321.51	10.92	1,310.59	0.00
		Total payments $158,580.88	Total interest paid $58,580.88	Total principal paid $100,000.00	

Ex.) Yearly Loan Payments

Year j	$(1+0.1)^{j-1}$ Column sum = 15.937425	Constant payment, $/yr, Eq. E	Interest payment, $/yr, Eq. B	Principal payment, $/yr, Eq. D	Remaining principal, $, Eq. C
0					100,000.00
1	1.000000	16,274.54	10,000.00	6,274.54	93,725.46
2	1.100000	16,274.54	9,372.55	6,901.99	86,823.47
3	1.210000	16,274.54	8,682.35	7,592.19	79,231.27
4	1.331000	16,274.54	7,923.13	8,351.41	70,879.86
5	1.464100	16,274.54	7,087.99	9,186.55	61,693.31
6	1.610510	16,274.54	6,169.33	10,105.21	51,588.10
7	1.771561	16,274.54	5,158.81	11,115.73	40,472.37
8	1.948717	16,274.54	4,047.24	12,227.30	28,245.07
9	2.143589	16,274.54	2,824.51	13,450.03	14,795.04
10	2.357948	16,274.54	1,479.50	14,795.04	0.00
		Total payments $162,745.39	Total interest paid $62,745.39	Total principal paid $100,000.00	

- In general, the more frequently loan payments are made, the less total interest is paid

돈의 시간적 가치 및 수익성

◆ 제14장. 생산비용 및 매출 추정 ▶ 345

Time Value of Money (1/2)

❖ **Money when invested, makes money**

- ✓ Money has a time value because it can earn more money over time
  - Money earned in the early years of the project is more valuable than that earned in later years

- ➢ Time value of money is measured in term of interest rate

Time Value of Money (2/2)

❖ Future worth

- ✓ Future worth of an amount of money, P, invested at interest rate, i, for n years is

$$\text{Future worth in year } n = P(1+i)^n$$

- ✓ Hence, the present value of a future sum is

$$\text{Present value of future sum} = \frac{\text{future worth in year } n}{(1+i)^n}$$

Net Present Value (NPV)

> NPV is the sum of the present value of the future cash flows

$$\mathrm{NPV} = \sum_{n=1}^{n=t} \frac{CF_n}{(1+i)^n}$$

where CF_n = cash flow in year n
t = project life in years
i = interest rate (= cost of capital, percent /100)

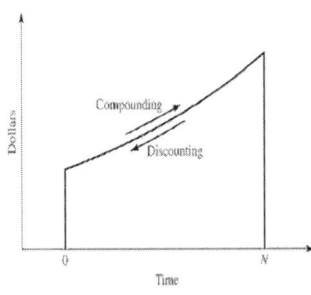

- Schematic diagram of compounding and discounting

NPV Key takeaways

✓ NPV is used to calculate the current value of a future streams of payments from a company, project, or investment

✓ To calculate NPV, we need to estimate the timing and amount of future cash flows and pick a discount rate equal to the minimum acceptable rate of return

✓ The discount rate may reflect our cost of capital or the returns available on alternative investments of comparable risk

✓ If the NPV is positive, it means its rates of return will be above the discount rate

NPV Formulas : Pros and Cons

Pros
- Considers the time value of money
- Incorporates discounted cash flow using a company's cost of capital
- Returns a single dollar value that is relatively easy to interpret
- May be easy to calculate when leveraging spreadsheets or financial calculators

Cons
- Relies heavily on inputs, estimates, and long-term projections
- Doesn't consider project size or return on investment (ROI)
- May be hard to calculate manually, especially for projects with many years of cash flow
- Is driven by quantitative inputs and does not consider nonfinancial metrics

Depreciation (감가상각)

> Depreciation is an accounting practice used to spread the cost of a tangible and physical asset over its useful life

- It represents how much of the asset's value has been used up in any given period
- Companies depreciate assets for both tax and accounting purposes

  - The value and worth of the plant decreases with time
  - Even if the equipment and facilities are well maintained, they become obsolete and of little value
  - When the plant closed, the plant equipment can be salvaged and sold only a fraction of the original cost
  - All fixed cost associated with new construction, but land cannot be depreciated

Types of Depreciation

➢ Straight-line method

➢ Declining Balance method

➢ Double Declining-Balance method (DDB)

➢ Modified accelerated cost recovery system (MACRS)

Recovery periods for chemical industry

Type of assets	Recovery period, years	
	MACRS	Straight line
Heavy general-purpose trucks	5	5
Industrial steam and electric generation and/or distribution systems	15	22
Information systems (e.g., computers)	5	5
Manufacture of chemicals and allied products (including petrochemicals)	5	9.5
Manufacture of electronic components, products, and systems	5	5
Manufacture of finished plastic products	7	11
Manufacture of other (than grain, sugar, and vegetable oils) food and kindred products	7	12
Manufacture of pulp and paper	7	13
Manufacture of rubber products	7	14
Manufacture of semiconductors	5	5
Petroleum refining	10	16
Pipeline transportation	15	22
Gas utility synthetic natural gas (SNG)	7	14
SNG—coal gasification	10	18
Liquefied natural gas plant	15	22
Waste reduction and resource recovery plant	7	10
Alternative energy property	5	12

†Source: © 2002 CCH Incorporated. All Rights Reserved. Reprinted with permission from *1997 Depreciation Guide Featuring MACRS*.

Straight-line method

> Straight-line is the most basic way to record depreciation

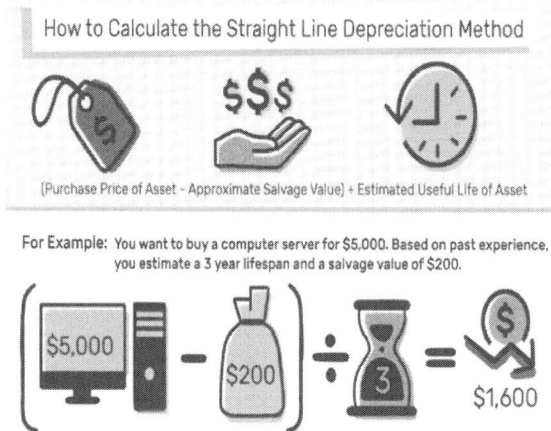

Source : www.investopedia.com/terms/d/depreciation.asp

Declining Balance(DB) & DDB method

> DB is an accelerated depreciation method
> - Because an asset's carrying value is higher in early years, the same percentage causes a larger depreciation expense amount

✓ Declining Balance Depreciation = Book Value x (1/Useful Life)

✓ DDB = Book Value x (2/Useful Life)

❖ If the machine costs $5,000 and has a useful life
  - using DB method, in year one, depreciation = $1,000
                          in year two, $800, and in year three, $640
  - using DDB method, in year one, depreciation = $2,000, in year two, $1,200

Source : www.investopedia.com/terms/d/depreciation.asp

MACRS method

> MACRS is used for most income tax purposes and for most economic evaluations

✓ MACRS method
- is based upon DDB method, but with no salvage or scrap value
- use the half-year convention
- can be switched to straight-line method, when SL method yields a greater depreciation allowance for that year

MACRS : Pros and Cons

> Pros
- allow to claim a larger depreciation expenses in the early years of an asset's life, which can provide significant tax savings
- relatively easy to use and provides a clear schedule of depreciation rates for different types of assets
- can be used to calculate depreciation for tax and financial reporting purposes

> Cons
- can result in a lower book value for an asset in the early years of its life, which can impact a company's financial statements
- is only available for certain types of assets, and the depreciation rates can vary depending on the asset's class and recovery period
- can be complicated to use for business that own multiple assets with different recovery periods

Source : fastercapital.com/content/Term--MACRS--Modified-Accelerated-Cost-Recovery-System.html

MACRS depreciation rates

Recovery year	Recovery period					
	3-year	5-year	7-year	10-year	15-year	20-year
	Depreciation rate, %					
1	33.33	20.00	14.29	10.00	5.00	3.750
2	44.45	32.00	24.49	18.00	9.50	7.219
3	14.81	19.20	17.49	14.40	8.55	6.677
4	7.41	11.52	12.49	11.52	7.70	6.177
5		11.52	8.93	9.22	6.93	5.713
6		5.76	8.92	7.37	6.23	5.285
7			8.93	6.55	5.90	4.888
8			4.46	6.55	5.90	4.522
9				6.56	5.91	4.462
10				6.55	5.90	4.461
11				3.28	5.91	4.462
12					5.90	4.461
13					5.91	4.462
14					5.90	4.461
15					5.91	4.462
16					2.95	4.461
17						4.462
18						4.461
19						4.462
20						4.461
21						2.231

General depreciation system
Applicable depreciation method: 200 or 150 percent
Declining balance switching to straight-line method
Applicable recovery periods: 3, 5, 7, 10, 15, 20 years
Applicable convention: half-year

*Source: © 2002 CCH Incorporated. All Rights Reserved. Reprinted with permission from 2000 U.S. Master Tax Guide.

Profitability evaluation

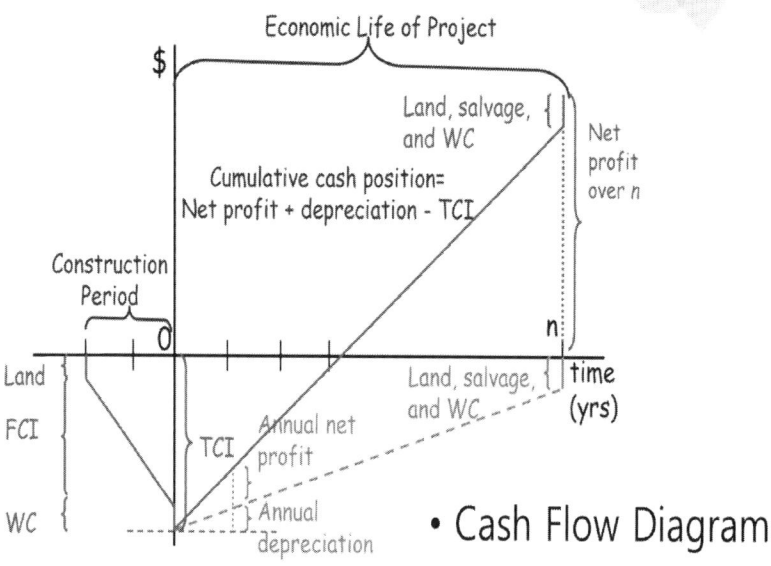

- Cash Flow Diagram

Methods for calculating Profitability

- ➢ Non-discounted methods
  (Do not consider the time value of money)

  - ● Return on Investment (ROI)
    - It is defined as the ratio of profit to investment

  - ● Payback Period (PBP)
    - It is defined the length of time necessary for the total return to equal the capital investment

- ➢ Discounted methods
  (consider the time value of money)

  - ● Net Present Value (NPV)
    - It is the difference between the present value of cash inflows and the present value of cash outflow over a periodical time

  - ● Internal Rate of Return (IRR)
    - It is a discount rate that makes the NPV of all cash flows equal to zero in a discounted cash flow analysis

Methods for calculating Profitability

- ➢ Non-discounted methods
  (Do not consider the time value of money)

  - ● Return on Investment (ROI)

    $$\mathrm{ROI} = \frac{\text{Current Value of Investment} - \text{Cost of Investment}}{\text{Cost of Investment}}$$

    - Current Value of Investment : The proceeds obtained from the sales of the investment of interest

  - ● Payback Period (PBP)

    $$\text{Payback Period} = \frac{\text{Cost of Investment}}{\text{Average Annual Cash Flow}}$$

Suggested Values for ROI

- Suggested values for risk and minimum acceptable return on investment

Investment description	Level of risk	Minimum acceptable return m_{ar} (after income taxes), percent/year
Basis: Safe corporate investment opportunities or cost of capital	Safe	4–8
New capacity with established corporate market position	Low	8–16
New product entering into established market, or new process technology	Medium	16–24
New product or process in a new application	High	24–32
Everything new, high R&D and marketing effort	Very high	32–48+

Methods for calculating Profitability

➢ Discounted methods
(consider the time value of money)

- Net Present Value (NPV)

$$NPV = \frac{\text{Cash flow}}{(1+i)^t} - \text{initial investment}$$

where:
i = Required return or discount rate
t = Number of time periods

- Internal Rate of Return (IRR)

$$0 = NPV = \sum_{t=1}^{T} \frac{C_t}{(1+IRR)^t} - C_0$$

where:
C_t = Net cash inflow during the period t
C_0 = Total initial investment costs
IRR = The internal rate of return
t = The number of time periods

Use of Profitability measures

- Use of Profitability measures[†]

Evaluation method	Percentage use	
	Small companies	Large companies
Payback period	43	52
Return on investment	22	34
Net present worth	16	80
Discounted cash flw rate of return	11	78

[†]E. J. Farragher, R. T. Kleiman, and A. P. Sahu, *Eng. Econ.*, **44**(2): 137 (1999).

- Net Present Worth = (the total of the present worth of all cash flow) − (the present worth of all capital investments)

Sensitivity for Economic Analysis

- The economic analysis of a project can only based on the best estimates that can be made of investment required and the cash flows

- The actual cash flows achieved will be affected by changes in raw materials costs and other operating costs, and will be very dependent on the sales volume and price

- Sensitivity analysis is a way of examining the effect of uncertainties in the forecasts on the viability of a project

Sensitivity Analysis Parameters

Parameter	Range of Variation
Sales price	±20% of base (larger for cyclic commodities)
Production rate	±20% of base
Feed cost	−10% to +30% of base
Fuel cost	−50% to +100% of base
Fixed costs	−20% to +100% of base
ISBL capital investment	−20% to +50% of base
OSBL capital investment	−20% to +50% of base
Construction time	−6 months to +2 years
Interest rate	base to base + 2 percentage points

Simple Method of Statistical Analysis

$$\text{mean value, } \bar{x} = \frac{(H + 2ML + L)}{4}$$

$$\text{standard deviation, } S_x = \frac{(H - L)}{2.65}$$

- ✓ Where, ML is most likely value, an upper value, H, and a lower value, L
- ✓ H and L can be estimated using the ranges of variation given in sensitivity parameters

제15장
Chemical Eng. Design Project
– A case study approach

Sensitivity for Economic Analysis

➢ The economic analysis of a project can be only based on the best estimates that can be made of investment required and the cash flows

➢ The actual cash flows achieved will be affected by changes in raw materials costs and other operating costs, and will be very dependent on the sales volume and price

➢ Sensitivity analysis is a way of examining the effect of uncertainties in the forecasts on the viability of a project

Sensitivity Analysis Parameters

Parameter	Range of Variation
Sales price	±20% of base (larger for cyclic commodities)
Production rate	±20% of base
Feed cost	−10% to +30% of base
Fuel cost	−50% to +100% of base
Fixed costs	−20% to +100% of base
ISBL capital investment	−20% to +50% of base
OSBL capital investment	−20% to +50% of base
Construction time	−6 months to +2 years
Interest rate	base to base + 2 percentage points

Simple Method of Statistical Analysis

$$\text{mean value, } \bar{x} = \frac{(H + 2ML + L)}{4}$$

$$\text{standard deviation, } S_x = \frac{(H - L)}{2.65}$$

- ✓ Where, ML is most likely value, an upper value, H, and a lower value, L
- ✓ H and L can be estimated using the ranges of variation given in sensitivity parameters

화학공정 설계 vs 화학제품 설계

| PROCESS DESIGN | PRODUCT DESIGN |

Today's Book Review

Topics in chemical engineering Volume 6

CHEMICAL ENGINEERING DESIGN PROJECT
A Case Study Approach
Martyn S. Ray and David W. Johnston

Topics in Chemical Engineering Series

Volume 1   HEAT AND MASS TRANSFER IN PACKED BEDS

Volume 2   THREE-PHASE CATALYTIC REACTORS

Volume 3   DRYING: PRINCIPLES, APPLICATIONS AND DESIGN

Volume 4   THE ANALYSIS OF CHEMICALLY REACTING SYSTEMS: A Stochastic Approach

Volume 5   CONTROL OF LIQUID-LIQUID EXTRACTION COLUMNS

Volume 6   CHEMICAL ENGINEERING DESIGN PROJECT
           . A Case Study Appidach

Published in 1989

About This book

- This book provides a case study approach the work involved in chemical engineering design project

- The approach adopted here is to provide brief notes and references for wide range of topics to be considered in the design project

- Case study material concerning *"The manufacture of Nitric Acid"* is presented, which is adapted from the design project performed by author, D.W. Johnston, in 1986

In This book

- Author's main advice to the student undertaking a chemical engineering design project is :
  'don't work in a vacuum!' : Do not assume that you are alone, or should, complete this project unaided

- May obtain information and help from many sources as you can find

- The completed project should be a testimonial to the student's abilities as a chemical engineer, soon to be employed in industry and eventually to become a recognized professional engineer

Contents

> **Part I : Preliminary Design**
> - Technical & Economic Feasibility

- ✓ Chap.1 The Design Problem
- ✓ Chap.2 Feasibility Study & Literature Survey
- ✓ Chap.3 Process Selection
- ✓ Chap.4 Process Description & Equipment List
- ✓ Chap.5 Site Consideration
- ✓ Chap.6 Economic Evaluation
- ✓ Chap.7 Mass & Energy Balances

> **Part II : Detailed Equipment Design (Case Study)**

- ✓ Chap.8 The Detailed Design Stage
- ✓ Chap.9 Absorption Column Design
- ✓ Chap.10 Steam Superheater Design
- ✓ Chap.11 Bleaching Column Design
- ✓ Chap.12 Nitric Acid Storage Tank Design

Chap.1 The Design Problem

1.1 Initial Consideration and Specification

- The following six steps in the design a chemical process have been identified

  1. Conception and definition
  2. Flowsheet development
  3. Design of equipment
  4. Economic analysis
  5. Optimization
  6. Reporting

- Feasibility study and Initial design considerations

  ✓ Product : Nitric acid (HNO_3)

  ✓ Capacity : 280 ton/day (92,400 ton/yr)

  ✓ 8,000 hours of operation (330days)

  ✓ Location : Bunbury, Western Australia
  (On this site, Ammonia and Ammonium nitrate plants are located)

  ✓ ~70% of product acid will be consumed in situ for ammonium nitrate

Chap.1 The Design Problem

1.2 Defining the Problem and Background Information

- **Introduction**
  - To determine both economically and technically feasible plant to produce HNO₃ in Western Australia

- **Properties and Uses**

- **The Evolution of HNO₃ Production Processes**

- **Ammonia Oxidation Chemistry**

(Chemical reactions for the oxidation of ammonia)

Main reactions	Heat of reaction (ΔH_r; kJ/mole)
1. $NH_3(g) + 2O_2(g) \rightleftharpoons HNO_3(aq) + H_2O(l)$	-436.918
2. $4NH_3(g) + 5O_2(g) \rightleftharpoons 4NO(g) + 6H_2O(l)$	-226.523
3. $2NO(g) + O_2(g) \rightleftharpoons 2NO_2(g)$	-57.108
4. $2NO_2(g) \rightleftharpoons N_2O_4(g)$	-28.617
5. $3N_2O_4(g) + 2H_2O(l) \rightleftharpoons 4HNO_3(aq) + 2NO(g)$	-15.747
6. $3NO_2(g) + H_2O(l) \rightleftharpoons 2HNO_3(aq) + NO(g)$	58.672

Chap.2 Feasibility Study & Literature Survey

2.1 Initial Feasibility Study

- At this stage, the process route has not yet been finalized although a preferred route may be apparent
- To obtain information regarding the alternative process routes, and to provide an assessment of the sustainability of these routes for a project

2.2 Presentation of Literature Surveys for Projects

2.3 Case Study – Feasibility Study (Market Assessment)
- Domestic and Global Market
- Market Analysis and Assessment Conclusions

2.4 Case Study – Literature Survey
- General Information (including Thermodynamic Data) and Process Technology
- Cost Estimation and Market Data

* The most used book for Cost Estimation is the
"Plant Design and Economics for Chemical Engineers"

Chap.3 Process Selection

3.1 Process Selection - Consideration

a) Will the process produce what the customer requires?
b) Is it possible to design, build and operate this plant economically?
c) The necessary design data, technology, fabrication and raw materials, etc., must be available
d) Plant must operate in a safe manner, providing an acceptable hazard risk to the employees and the public
e) Plant must conform to any environmental protection requirements, and any possible restrictions
f) Plant must be as energy efficient (and energy self-sufficient) as possible
g) Maintenance requirements should be minimized
h) Plant should be designed to operate adequately under conditions of reduced throughout (by 50%), and for increased production (~25%)
i) The production of any unusable by-products should be minimized
j) All necessary utilities should be available

3.2 Case Study – Process Selection

- Process Comparison : Factors for Dual-Pressure Process vs Single-Pressure Process
- Other Considerations : Utilities, Space, Capacity, and, etc.
- Process Selection Conclusion : Single-Pressure Process

Chap.4 Process Description & Equip. List

4.1 Introductory Notes

a) Specific type of equipment
b) Size and/or capacity
c) Material of construction
d) Operating Pressure
e) Maximum operating temperature, or minimum temperature
f) Insulation required
g) Corrosion allowance (if large)
h) Special features, e.g. jackets on heat exchangers
i) Duplication of plant items (for safety and/or reliability)

4.2 Case Study – Process Description

- Process introduction
- Requirements of Major Process Units
- Mechanical Design Features of Major Units
- Process Flow Diagram & Process Performance Assessment

PFD

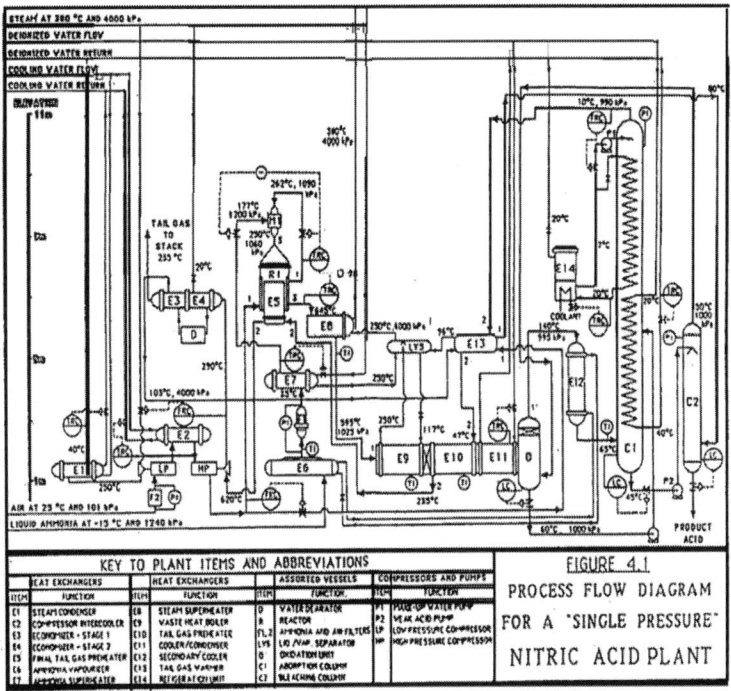

FIGURE 4.1 PROCESS FLOW DIAGRAM FOR A "SINGLE PRESSURE" NITRIC ACID PLANT

Equipment Description

Unit	Function
Ammonia storage vessel	500m3 for one week supply of feed.
Ion-exchange unit	Provision of deionized water.
Deionized water cooler	Cool circulating deionized water.
Air filter	Remove solid particulates from feed air.
Two-stage air compressor	Provide feed air at 1090 kPa.
Ammonia vaporizer	Evaporate liquid ammonia feed.
Ammonia filter	Remove solid particulates from feed.
Ammonia superheater	Heat ammonia vapour to approx. 180°C.
Feed mixer	Mix the two gas-feed streams.
Reactor	Perform the oxidation of ammonia in air.
Waste-heat boiler	Cool reaction gases and produce steam.
Vapour/liquid separator	Remove entrained liquid as feed for the waste-heat boiler.
Steam superheater	Cool reaction gases and superheat steam.
Tail-gas preheater	Cool reaction gases and preheat tail gas.
Tail-gas warmer	Provide first stage of tail gas preheat.
Cooler/condenser	Condense out weak acid.
Oxidation unit	Final oxidation of reaction gases.
Secondary cooler	Cool reaction gases prior to absorption.
Refrigeration unit	Cool make-up water and provide some cooling.
Absorber	Absorb nitrogen oxides to form nitric acid.
Bleaching column	Remove dissolved NOx to give product acid.
Nitric acid product tank	2000m³ storage of one week supply of product.

Chap.5 Site Considerations

5.1 Site Selection

5.2 Plant Layout

5.3 Environmental Impact Analysis

5.4 Case Study – Site Considerations
- Perth Metropolitan Region
- Country Districts
- Site Location Conclusion and Plant Layout
- Environmental Impact Analysis

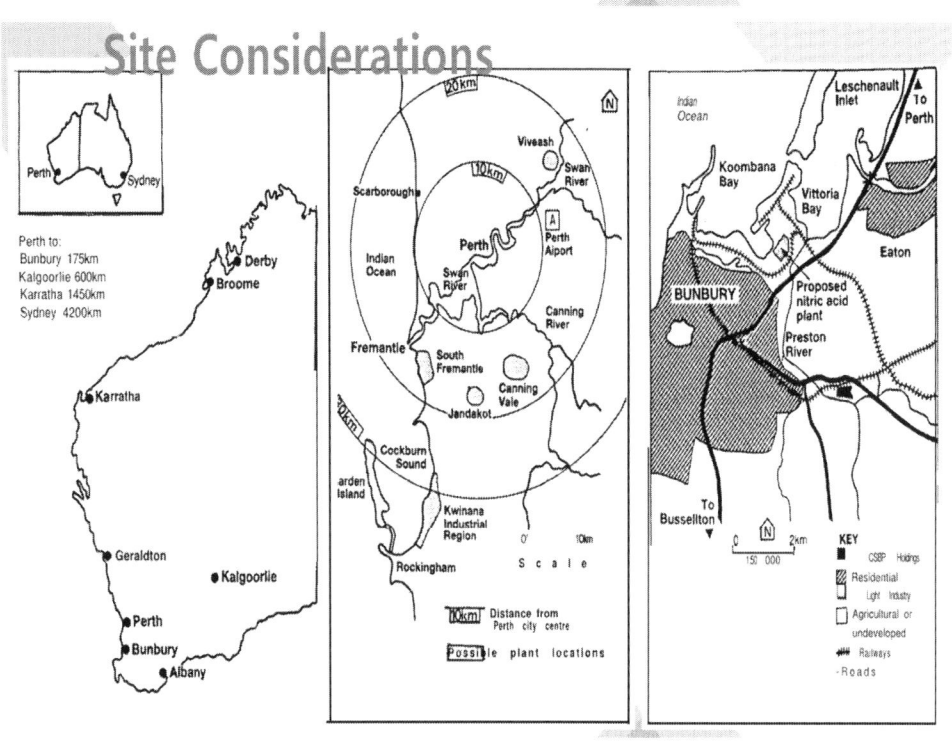

Overall Layout

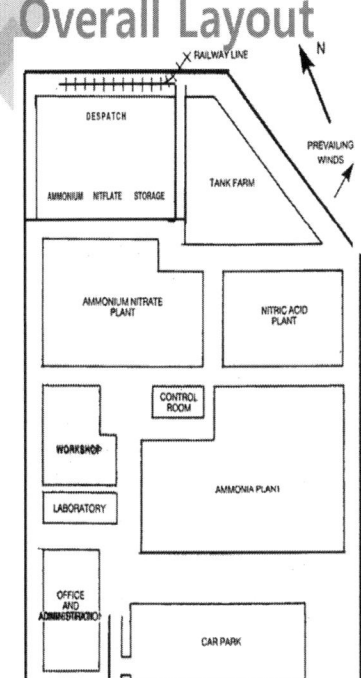

Plant Layout

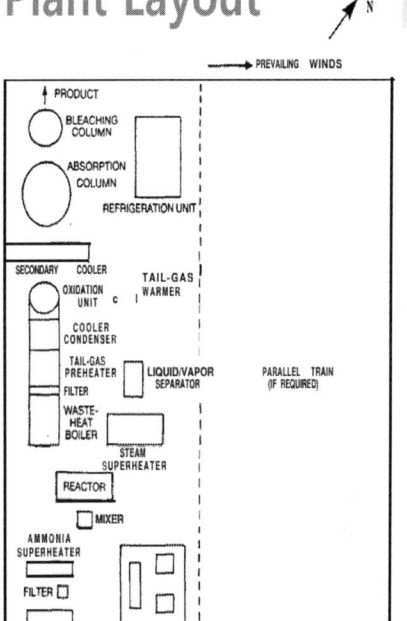

Chap.6 Economic Evaluation

6.1 Introductory Notes
- The economic evaluation of chemical plant covers a wide range of topics
- Several excellent reference sources are available, those written for chemical engineers are useful
* The most prominent text in this filed is **"Plant Design and Economics for Chemical Engineers"**

6.2 Capital Cost Estimation
- Cost of Equipment (Major Items) : Six-tenth rule & Lang factor method
- Auxiliary Services

6.3 Operating Costs

6.4 Profitability Analysis
- ROI (Return on Investment), PBP (Payback Period)

6.5 Case Study – Economic Evaluation
- Capital Cost Estimation
- Investment Return

Key review item for economic evaluation
(In the book of "Plant Design and Economics for Chemical Engineers")

- ➢ Cost and Asset Accounting
  - presents the analysis od costs and profits as used for industrial operations
- ➢ Cost Estimation
  - provides information regarding the fixed capital investment and operating expenditure
- ➢ Interest and Investment Costs
  - discusses the concept and calculation of investment (payment) for the use of borrowed capital
- ➢ Taxes and Insurances
  - Taxes represents a significant payment from a company's earnings
- ➢ Depreciation
  - measures the decrease in value of an item with respect to time
- ➢ Profitability
  - measures the amount of the profits generated

Chap.7 Mass & Energy Balances

7.1 Preparation of Mass and Energy Balances : should includes the following details

a) Mass flow of all stream into and out of the equipment per unit time (molar or volume flowrate must not used)
b) Composition (mass %) of all stream
c) Sometimes the molar flowrate (and mole %) of gas stream are also included
d) Temperature of each stream (use °C or K)
e) Pressure of each stream (use bar or kg/cm²)
f) Enthalpy content of each stream (J, MJ, GJ, etc.)

* Mass balances often do not balance exactly, some error is allowable (1%, 5%, 10%?)

7.2 Preliminary Equipment Design

- The preliminary design reassess and modify the data, and assumptions made in earlier stage of the project

7.3 Computer-Aided Design

7.4 Case Study – Mass and Energy Balanced

- Overall Process Mass Balance
- Unit Mass and Energy Balances

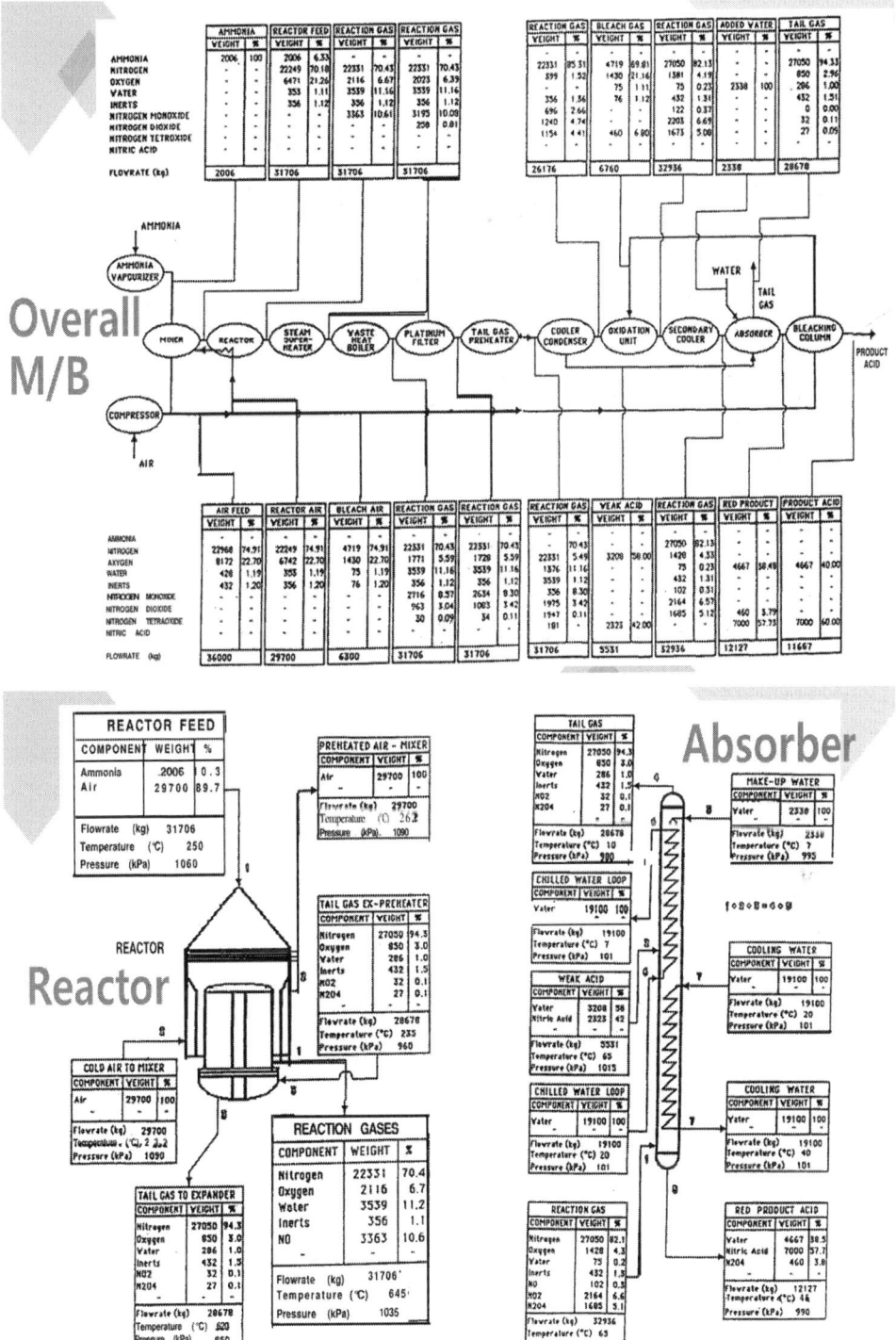

저탄소 플랜트 개념설정과
경제성 및 사업타당성 분석

2024년 12월 10일 초판 1쇄 인쇄
2024년 12월 15일 초판 1쇄 발행

저　자	고 동 현 ・ 著
발 행 처	도서출판 에듀컨텐츠휴피아
발 행 인	李 相 烈
등록번호	제2017-000042호 (2002년 1월 9일 신고등록)
주　소	서울 광진구 자양로 28길 98, 동양빌딩
전　화	(02) 443-6366
팩　스	(02) 443-6376
e-mail	iknowledge@naver.com
web	http://cafe.naver.com/eduhuepia
만든사람들	기획・김수아 / 책임편집・이진훈 정민경 하지수 박현경 황수정 디자인・유충현 / 영업・이순우

ISBN　　978-89-6356-491-3 (93530)
정　가　　26,000원

ⓒ 2024, 고동현, 도서출판 에듀컨텐츠휴피아

> 이 책은 저작권법에 따라 보호받는 저작물이므로 무단전재와 무단복제를 금지하며, 책 내용의 전부 또는 일부를 이용하려면 반드시 저작권자 및 도서출판 에듀컨텐츠휴피아의 서면 동의를 받아야 합니다.